INTERCONNECTION NOISE IN VLSI CIRCUITS

Interconnection Noise in VLSI Circuits

by

Francesc Moll
Universitat Politècnica de Catalunya,
Barcelona, Spain

and

Miquel Roca
Universitat Illes Balears,
Palma de Mallorca, Spain

KLUWER ACADEMIC PUBLISHERS
BOSTON / DORDRECHT / LONDON

A C.I.P. Catalogue record for this book is available from the Library of Congress.

ISBN 1-4020-7733-5

Published by Kluwer Academic Publishers,
P.O. Box 17, 3300 AA Dordrecht, The Netherlands.

Sold and distributed in North, Central and South America
by Kluwer Academic Publishers,
101 Philip Drive, Norwell, MA 02061, U.S.A.

In all other countries, sold and distributed
by Kluwer Academic Publishers,
P.O. Box 322, 3300 AH Dordrecht, The Netherlands.

Printed on acid-free paper

Printed in the Netherlands.

Contents

List of Figures

List of Tables

To our families

Preface

Interconnections are a most important design issue nowadays. Trends in the microelectronic industry are leading to unwanted interconnect effects, especially noise, becoming more important. This increasing importance is mainly due to three reasons: increasing integration, increasing signal frequency spectrum components, and increasing complexity.

The increase in integration and frequency of signals account for coupling problems between adjacent lines and a growing importance of parasitic components (capacitance and inductance). These two phenomena introduce analog effects in digital design, and are therefore direct causes of the noise problem.

The increase in complexity is perhaps more indirectly related to noise and the interconnection problem, but it is also very important. The drive to reduce the time to market of new electronic products make design verification extremely important and this pre-fabrication verification must be as accurate as possible to reduce the risk of having to redesign failed prototypes. However, the analog effects introduced by interconnections make traditional digital verification tools inappropriate for addressing the problem. Recently there have been many advances in interconnect simulation algorithms and efficient simulators can calculate solutions for sophisticated models. This is a very important subject and, with these algorithms and computer availability, accurate noise waveforms for a small number of coupled interconnections can nowadays be calculated easily. The problem is that from the point of view of the whole integrated system, the applicability of these sophisticated models is necessarily limited because today's digital designs are so complex and the number of interconnections is so large that a complete electrical simulation of the whole chip is impossible, as it would take weeks or months.

Given this complexity problem for verification, there are two possible solutions: one is to simplify the interconnect models. The other is to address interconnect issues from the beginning of the design process, so they can be in some way implemented as design rules in the design flow.

With respect to interconnect models, the old consideration of an interconnect as an ideal short-circuit has successively evolved to a capacitive, a resistive-capacitive, and finally an inductive model and transmission line models, as the necessity to include more effects has arisen. However, the fundamental basis of these models is often unknown to electrical designers, so it is difficult to judge when one model or another is more suitable.

The establishment of a design flow oriented to integrate interconnect problems must start from the knowledge of what the mechanisms for these noise problems are. Eventually, this knowledge must be implemented into CAD tools. Some existing physical design tools already take into account these effects. Other tools are needed, such as functional verification tools modeling interconnection noise as high-level models, and expert systems that automatically guide the designer.

This book addresses two main problems with interconnections at the chip and package level: crosstalk and simultaneous switching noise. It is intended to provide the notions required for understanding the problem of modeling starting from physical arguments, so that it is possible to select an appropriate interconnection model that is both simple and accurate for the type of problems arising. Also, simple models of crosstalk and switching noise are used to give general ideas that may eventually lead to CAD implementation as discussed above. Finally, some verification and test issues related to interconnection noise are discussed. Throughout the book, the examples used to illustrate the discussion are based on digital CMOS circuits, but the general treatment of the problems is made from a fundamental point of view, so that the discussion can be applied to different technologies. The book should be of interest to chip designers, especially to digital designers dealing with interconnect problems who want a deeper explanation of these phenomena. In this sense, the book's orientation is towards giving general information rather than being a compilation of practical cases. Each chapter contains a list of references for the topics dealt with, both recent and classic ones.

The six chapters of the book are grouped in the three topics mentioned above. The first two chapters provide a physical foundation for on-chip and package interconnect models. First, chapter 1 is an introduction describing the general aspects of the technology of this type of interconnection, while the main topic of chapter 2 is the derivation of electrical models of interconnections and their relationship to electromagnetic wave propagation laws. In the second block, chapters 3 and 4 describe and model the kind of noise effects of on-chip and package interconnections respectively. The third block relies on the previous chapters to derive some general design rules regarding on-chip and package interconnections in 5, and finally, chapter 6 deals with a not so familiar topic: detection and testing for interconnection-induced noise.

Acknowledgments

During the preparation of this manuscript, there were many people who made the work possible.

Firstly, we would like to thank professor Antonio Rubio for introducing us to this subject at the beginning of our scientific careers. His encouragement and enthusiasm always guided us. We also want to thank our respective research groups at the UPC in Barcelona and at the UIB in Balearic Islands. At the UIB, we particularly would like to thank Eugeni Isern for his collaboration on part of the results, and for his help with the editing tasks.

Also, Francisco Mesa, from the University of Sevilla, kindly answered some questions on wave propagation in lossy media.

Finally, our families were very patient while the manuscript was being prepared. Thanks to our wives, Teresa and Maria, and daughters, Maria and Antonia.

This work has been partially supported by the Spanish Ministry of Science and Technology, and the European Regional Development Fund (ERDF) from the European Union through the projects TIC2001-2337 and TIC2002-1238.

Chapter 1

MICROELECTRONIC INTERCONNECTIONS

The research topic known as Signal Integrity, or also, as recently proposed, Electrical Modeling of Interconnects and Packages [1], is a very important part of the electronic system design process. Its objective is to obtain a model of the physical phenomenon of signal propagation that can be used by electronic designers to predict any distortion of the information transmitted. Thus, the extracted model can either be electrical, for use with electrical design using currents and voltages as observable variables, or it can be logical, for use with higher level description tools.

The problems concerning signal propagation through interconnects are briefly explained in the first section. After this, the main technological characteristics of microelectronic interconnections, inside the chip and at package level, are also explained.

1. Signal transmission in interconnects

Any electronic system, at any level, is composed of functional blocks (transistors, gates, sub-circuits, IPs, cores, processors, boards...) interconnected with each other; that is, capable of transmitting information from one part to another. This information is in the form of a voltage or a current value.

Ideally, the communication between two interconnected blocks should be instantaneous and without any distortion. Under these conditions, the system is completely described by its building blocks and the way they are connected. However, this ideal picture cannot be achieved in practice. The reason is that physically, there is always a propagation time for transmitting the information from one point to the other. If the signals vary very slowly compared to this propagation time, the transmission may be considered instantaneous and the ideal picture is valid. If, on the other hand, the signals vary rapidly compared to the propagation time, several effects may be observed:

- Signal Delay. This fact must be taken into account when designing the system to ensure that all blocks receive the information at the expected time.

- Reflections coming from the end of the interconnection and interfering with the signal sent.

- Interference between nearby interconnections (crosstalk).

All these non-ideal effects may influence the behavior of the system and they represent a limitation on performance. If not properly addressed at the design stage, they may be large enough to completely distort the transmitted information and therefore cause a system malfunction.

1.1 The problem with interconnections

The evolution of electronic technology, as dictated by Moore's law, is towards denser, more complex, and faster systems. All three trends imply a larger interconnection effect, once considered only from the functional point of view. Today, a microelectronic design that does not consider interconnection effects from the beginning of the design process is most probably condemned to a large number of trial and error iterations, with the associated increase in cost and time to market.

The reason for the ever-increasing importance of interconnections can be understood using electrical models of the interconnections. The physical foundations of the electrical models will be fully justified in Chapter 2, but let us for the moment use them as an introductory explanation.

The electrical parameters that characterize an interconnection are resistance (R), capacitance (C) and inductance (L). The interconnection response to a certain voltage or current excitation can be derived from the relative values of those parameters. Roughly speaking, one can consider that the delay is mostly dependent on R and C (although L is also important for high frequency signals). On the other hand, crosstalk interference depends on mutual C and L between close lines. Let us qualitatively analyze the influence of technology trends on parameters and their effect on signal integrity.

1.1.1 Increase in integration

As the number of devices per unit area increases, the interconnections between them must evolve towards a reduced cross-sectional dimension and a greater number of vertical conductor layers.

The reducing cross-sectional dimension leads to more tightly coupled interconnections, and therefore a higher probability of unwanted crosstalk interference between them.

The increased number of vertical interconnection layers imply that the upper layers are further away from the reference (connected to the substrate) unless neighboring upper layer conductors are devoted to reference. In general, though, upper layers show lower capacitance, and higher inductance, and this makes the conductors more susceptible to perturbations as higher frequency components in the signal spectrum become more important in new technologies.

At the same time, the increase in device density forces a reduction in cross-sectional dimensions that increases R. This increase in R produces a double effect: on one hand, adding to the delay budget, and on the other hand, producing a signal distortion.

1.1.2 Increase in complexity

The impact of chip technology trends on the delay was studied in the 1980s [2], [3], and it was shown that the increase in complexity implies an increase in the length of global interconnections, and therefore, in delay. This increase in length can be expressed as a function of the chip's area for global interconnections:

$$l \propto \sqrt{A} \tag{1.1}$$

Also associated with the increase in complexity, is an increase in number of conductor layers to interconnect the high number of devices.

1.1.3 Signal frequency

Device (transistor) driving capability increases as a result of technology scaling. This means higher frequency components in signals are transmitted through interconnections. Here, frequency refers not only to sinusoidal signals, but to the frequency spectrum of digital signals. From Fourier analysis, a digital signal with a smaller rise time, will show higher frequency components. A simplified analysis [4] can give a dependence between the rise/fall time of a digital signal transition and its maximum significant frequency, F_{knee}:

$$F_{knee} \approx \frac{0.5}{t_r} \tag{1.2}$$

Higher frequency content means two things: firstly, circuit parameters L and C will have a great influence on the line impedance (its effect is proportional to frequency). For example, coupling between lines will increase from DC or low-frequency signals to high frequency signal propagation, because the capacitance between them cause the impedance between one line and the other to reduce with frequency.

Secondly, reflections of the signal are greater for high-frequency signals. There are two ways of looking at this: one is in the time domain, considering that reflections will be important if the signal at the driver end has reached its full amplitude before arriving at the end of the line:

$$t_r \ll t_p \tag{1.3}$$

where t_p is the time of propagation of the signal along the line and t_r is the rise time of the signal at the driver end.

A second way of examining the importance of reflections is to relate the time of propagation to the line length (through propagation velocity), and the signal rise time to signal wavelength. Both magnitudes can then be compared by using a worst-case approach with the shortest wavelength, corresponding to F_{knee}:

$$l \gg \lambda_{knee} \tag{1.4}$$

1.2 Physical factors in interconnect design

As discussed above, the quality of signal transmission will eventually depend on physical factors of the interconnections that determine the respective values of R, C and L.

There are three physical factors influencing electrical parameters:

- Material properties of the conductors forming the interconnections and the dielectrics between them: conductivity in the case of conductors, and dielectric constant (as well as possible losses) from the dielectrics. This influences R and C respectively.
- Dimensions of conductor and dielectric materials: length, width and thickness of conductors, distance between centers of lines (pitch), and dielectric thickness, which determine distance between stacked interconnections. This influences R, C and L.
- Connection of the conductors carrying reference voltages, usually known as power supply distribution. Influence on C and L.

The first item depends exclusively on technology, the second one depends on technological restrictions and design, and the third item depends mostly on design, but is also restricted by the type of technology used.

2. On-Chip interconnections

2.1 Technology

A typical microelectronic manufacturing process consists of successive steps to create the transistors on a semiconductor substrate. After that, several layers of conductive material separated by dielectric layers are deposited and selectively etched to form the interconnections between transistors [5]. After the deposition process of the conductive material, a photolithography process defines the horizontal dimensions (width and pitch) of interconnections.

2.1.1 Conductive materials

Aluminum and copper are the two conductive materials currently used for on-chip interconnections. Each material defines the type of fabrication process: aluminum-based alloys for mature processes, and copper-based alloys for modern, high performance processes. The aluminum is deposited using a physical vapor deposition (PVD) or chemical vapor deposition (CVD) process. In addition, diffusion barriers based on TiN are deposited to avoid contamination of lower layers. The main limitation of the aluminum-based process is its higher resistivity, compared to copper. To have interconnection lines both narrow and with low resistance, the solution is to increase the metal thickness, or aspect ratio. This increase in aspect ratio makes the additional planarization process —either chemical, mechanical, or both— necessary. Aluminum can be doped with copper to obtain a lower resistivity interconnection.

The transition to copper interconnections, commercially introduced by IBM in the late 1990's, allows the manufacture of narrower lines with the same resistance as aluminum lines, and, therefore, more densely interconnected systems. The copper can be deposited by an electrochemical process, or by vapor deposition (CVD or PVD). Copper is a more serious contaminant than aluminum, and therefore, stop barriers based on tantalum are deposited before the copper to protect the underlying silicon devices. Also, the copper is not as easily etched as aluminum, so the damascene process is used, consisting of depositing the copper on previously etched oxide trenches, and later removing the upper copper layer by CMP (chemical mechanical polishing). This process uses an extra etch-stop layer, usually Si_3N_4 [6]. As this etch-stop layer has a high dielectric constant (around 5), its presence can influence the capacitance of the interconnection lines if it is not properly controlled.

The manufacturing difficulties related to Cu interconnections make it a more expensive solution than Al-based interconnections, but this extra cost can be compensated for by better performance. A hybrid alternative in which the lower layers are Al interconnections and the upper layers are Cu, is also possible. In either case, the effective resistivity of manufactured lines is higher than the bulk material's resistivity (Al or Cu) because the deposition process causes the formation of grains that affect electron mobility, and this fact must be taken into account when calculating line resistance.

2.1.2 Dielectric materials

The most common dielectric material between conductive layers is silicon dioxide (SiO_2) due to its good dielectric properties and its compatibility with the rest of the process steps. The dielectric material influences the value of capacitance between conductors through its dielectric constant (k). The higher the dielectric constant, the higher the line capacitance, and therefore, the bigger the delay and potential crosstalk interference between lines. With regard to delay in particular, there is a very intensive search for a compatible dielectric with a smaller dielectric constant (low-k dielectric materials). Silicon dioxide has a value of k around 3.9. New materials and deposition processes give rise to less dense materials that present constants between 2.5 and 1.5 [7], [8], although most of them are currently at the development stage.

2.2 Scaling trends

After the manufacturing process is finished, the physical structure of on-chip interconnections is the one represented in figure 1.1. High- performance processes nowadays offer up to seven or eight metal layers, while a low-cost process may have only two or three. The large number of metal layers allows the design of more compactly interconnected systems. Lower layers are devoted to local routing, interconnecting nearby elements (transistors inside gates and neighboring gates). The upper layers are devoted to global routing, transmitting signals along the whole chip and voltage references. As the global routing lines are the longest, their cross-section must be bigger than local routing lines, in order to keep their resistance into reasonable values. Therefore, a reverse scaling scheme is applied with respect to interconnections: the lower layers are scaled down with the rest of the technology, to increase interconnection density, while the upper layers are scaled up to decrease the resistance of global routing, as is schematically shown in figure 1.1.

2.3 Electrical reference connection schemes

The voltage signals transmitted by interconnections are actually a voltage difference with respect to a fixed voltage reference. The two references that are present in the electrical connection system are the power supply connections, denoted as V_{CC} or V_{DD} (highest voltage) and V_{SS} or GND, "ground"[1] (lowest voltage). The distance between the signal conductor and the conductors carrying the reference voltages has a severe influence on the quality of the signal waveform, improving this quality as the distance between signal and reference decreases, as will be seen in subsequent chapters. Thus, the way the reference voltages are physically distributed along the chip will also have an influence on the quality of the signals transmitted.

In silicon integrated circuits, the semiconductor substrate is connected to GND or V_{DD} through chip lines contacting the substrate in order to bias the

[1]Actually, the term "ground" used as the lowest voltage reference is misleading, because its actual voltage is not necessarily the same as the *real* ground voltage. However, as the term is used extensively, it will be employed here as equivalent to V_{SS}.

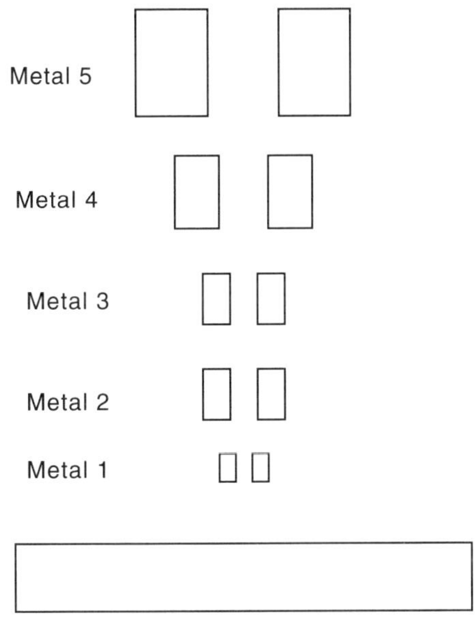

Figure 1.1. Cross section of a typical on-chip interconnection stack-up.

transistors. Therefore, the substrate can be considered as another reference conductor, sometimes being the closest reference conductor especially for lower layers. This means that the line capacitance of lower interconnect layers is very much dependent on the distance to substrate. Upper layers are mostly screened from the substrate effect by lower layers and, to ensure good signal integrity, reference lines must be routed close to them, especially for high-speed signals, which are more susceptible to noise perturbations.

The semiconductor substrate usually shows back metallization that makes electrical contact with the package die attach pad. This die attach pad can either be externally connected to the ground terminal or left unconnected. If connected, it can act as a small reference plane for lead-frame packages which do not provide special package layers for reference planes, and this can therefore improve signal transmission.

3. Package level interconnections

3.1 Technology

The manufacturing process and materials for microelectronic packages are more diverse than the chip manufacturing. Several characteristics of package families are summarized in chapter 4.

3.2 Scaling trends in packaging

Chips cannot accomplish the function for which they were designed if they cannot communicate with other components of the system sending and receiving data and receiving the voltage supply. Packages, then, have a double mission: to provide the physical substrate for communication and power supply and to give protection to the chip in hazardous environments. Another very important aspect is thermal management; the package must be able to provide a low thermal resistance path between the chip and the environment, such that the power dissipated by the chip does not increase its temperature excessively.

Packages provide the bridge between the silicon and system (board) level. As the technological development of both parts is not moving at the same pace, an increasing distance between them is observed. On one hand, the increase in chip complexity has translated as an increase in I/O demands. This increase is modeled as Rent's Rule [9], which gives an empirical formula relating the number of components in a given sub-system to the number of I/O terminals necessary for communication with the rest of the system:

$$T = K \cdot G^p \tag{1.5}$$

T being the number of I/O ports, K the mean I/O ports of each gate in the circuit, G the number of gates in the circuit, and p the so-called Rent exponent, an empirical parameter taking a value between 0.5 and 1 depending on the circuit architecture. This ratio expresses an exponential increase of I/O ports with circuit complexity. This is a tough challenge for circuit packaging, which must be capable of accommodating this large number of I/O ports and distributing them to the board level connections, which have a larger pitch than available for chip-level interconnections.

A simultaneous trend is to reduce the size and weight of the whole system, which obviously implies a need for the miniaturization of microelectronic packages. Chip Scale Packages (CSP) are packages that show a footprint area of at most 1.2 times that of the bare die [10]. However, the miniaturization possibilities for the package are not the same as for the chip.

At the same time, the package must propagate the signals without introducing more distortion. This is done using several manufacturing technologies and package types, from single layer interconnection structures to complex multi-layer interconnections, including reference planes at package level. The choice between the many alternatives is a complex one, where factors as diverse as signal integrity, mechanical reliability, power dissipation capability, size/weight, and cost must be taken into account.

3.3 Power voltage distribution and switching noise

The reference voltages are distributed in the package either through package leads, just as signal lines are, or through planes. The way they are distributed is an important issue affecting the quality of the signals transmitted. Some package technologies allow the use of planes for power distribution and some do not. In any case, a number of leads in the package will be devoted to power distribution and the way these leads are geometrically assigned has an influence on signal integrity, as will be shown in chapter 4.

Switching noise (or Simultaneous Switching Noise, SSN) is one of the main electrical effects introduced by packages. It is produced when there is a sudden current demand by a part of the circuit. Due to the inductance of the power and ground voltage distribution, the current derivative causes a voltage fluctuation to appear at the power and ground terminals of the circuit. This fluctuation affects not only the switching element, but the whole circuit and, if important enough, it may affect the logic value of storage elements and propagate to other interconnected chips.

As a first approximation, the noise can be expressed as a function of the number of switching elements of the circuit [11], [12]:

$$v_n = N L_{eff} \frac{dI}{dt} \tag{1.6}$$

being N the number of switching elements, and L_{eff} the so-called effective inductance. The value of the effective inductance is a combination of the self and mutual inductances of the different pins involved in the switching and can be put explicitly as a function of these inductances only for very simplified cases [12].

In general, the effective inductance concept is a particular case of the equivalent complex impedance of the power distribution system. For an ideal power distribution, its impedance should be zero, and in that case, there would not be any switching noise. In reality, there is complex impedance depending on frequency. When there is a current demand, a voltage is produced that alters the value of power supply voltage seen by the switching element, or by any other element connected to the same power distribution system. Thus, it is possible to characterize the power distribution by its equivalent impedance, as a function of frequency.

References

[1] Ruehli A.E. and Cangellaris A.C. "Progress in the methodologies for the electrical modeling of interconnects and electronic packages". *Proceedings of the IEEE*, 89(5), May 2001.

[2] Saraswat K. C. and Mohammadi F. "Effect of Scaling of Interconnections on the Time Delay of VLSI Circuits". *IEEE Journal of Solid State Circuits*, 17(2), April 1982.

[3] Bakoglu H. B. *Circuits, Interconnections and Packaging for VLSI*. Addison-Wesley, 1990.

[4] Johnson B., Quarles T., Newton A. R., Pederson D. O., and Sangiovanni-Vicentelli A. *SPICE3 Version 3e User's Manual*. University of California, Dpt. of Electrical Engineering and Computer Sciences. University of California. Berkeley, CA, 94720, 1991.

[5] Quirk M. and Serda J. *Semiconductor manufacturing technology*. Prentice Hall, 2000.

[6] Lin X.W. and Pramanik D. "Future interconnect technologies and copper metallization". *Solid State Technology*, 41(10), October 1998.

[7] Kikkawa T. "Present status and future trend of low-k dielectrics/interconnect technology for ULSI". In *7th International Symposium on Plasma- and Process-Induced Damage*, 2002.

[8] Lin S., Changming Jin, L. Lui, Minghsing Tsai, Daniels M., Gonzalez A., Wetzel J.T., Monnig K.A., Winebarger P.A., Jang S., Yu D., and Liang M.S. "Low-k dielectrics characterization for damascene integration". In *IEEE 2001 International Interconnect Technology Conference*, 2001.

[9] Donath W. E. "Wire Length Distribution for Placements of Computer Logic". *IBM J. Res. Develop.*, 25(3), May 1981.

[10] Lai Z. and Liu J. The nordic electronics packaging guideline. http://extra.ivf.se/ngl/, 2000.

[11] Davidson E. E. "Electrical Design of a High Speed Computer Package". *IBM J. Res. Develop.*, 26(3), May 1982.

[12] Katopis G. A. "Delta–I Noise Specification for a High–Performance Computing Machine". *Proceedings of the IEEE*, 73(9), September 1985.

Chapter 2

INTERCONNECT MODELING IN CIRCUIT SIMULATION

As stated in the previous chapter, the performance of electronic systems is more and more critically dependent on the effects introduced by interconnections. It is therefore very important to have good models of the interconnection effect at the design phase to prevent design-manufacturing iterations and to reduce the time to market of electronic products.

There are several possible models, with varying degrees of accuracy: from a simple capacitor to a frequency-dependent parameter transmission line. All of them are a more or less simplified interpretation or model of the fundamental physical phenomenon, which is the propagation of an electromagnetic wave in the complex metal-dielectric structure formed by the interconnection network of an electronic circuit. The problem is that a detailed electromagnetic analysis of a complete digital circuit with millions of interconnections and devices is beyond the computation capabilities possible today. Electromagnetic phenomena are therefore reduced to electrical models. Even complex electrical models are reduced, when accuracy is not critical, to simpler electrical models because the simulation time of all the system would otherwise be prohibitively long. The question is, then, how to assign a proper model to each interconnection sub- system so that the model predicts only the number of possible effects important for that sub-system, without being so complex that the simulation time is too large.

This chapter presents the different models, starting from a description of the fundamental physics of signal propagation, to give an understanding of the difference between them, and attempting to establish rules for selecting, a priori, the model to use when simulating a system.

1. Physical foundations for circuit models of interconnections

We will start by trying to understand the physics behind simple digital switching of a gate from the point of view of signal propagation in an interconnect.

Let us consider an idealized 2D physical interconnect structure, but still fundamentally similar to the phenomenon we are interested in, shown in figure 2.1. It consists of two conductors with an externally applied electric field between them. These two conductors represent the V_{DD} and V_{SS} lines in a circuit and the electric field is applied via an external power supply. In the middle of these conductors, a third conductor (what we will call a signal line) can be alternatively connected at one of its ends either to V_{DD} or to V_{SS} by two ideal switches (a model of a digital gate).

Figure 2.1 schematically shows the two different static charge and electric field distributions resulting from connecting the signal line either to one or the other reference metal. In one case, the electric field is confined in the region between the lower metal and the line, while it is null in the region between the signal and the upper metal. In the other case, the field is confined in the region close to the upper metal. Clearly then, a change of state implies a change in the electric field, which in turn causes the appearance of a magnetic field as described by Maxwell's equations:

$$\nabla \times \mathbf{E} = -\mu \frac{\partial \mathbf{H}}{\partial t} \tag{2.1}$$

$$\nabla \times \mathbf{H} = \mathbf{J} + \epsilon \frac{\partial \mathbf{E}}{\partial t} \tag{2.2}$$

An electromagnetic wave is therefore generated because of the switching event, propagated in the dielectric between conductors. This wave carries the information that the switching event occurred and will arrive at the end of the line, to be interpreted by the subsequent gate. At the end of the line it will probably be partially reflected, and, after some trips back and forth, it will eventually disappear, giving rise to a new static electric field distribution.

From the above description, then, it follows that an electronic circuit interconnection is in reality a waveguide: a metal structure to confine and guide the propagation of the electromagnetic signal. Its success in doing so will determine the quality of the interconnection and the signal it transmits (generally referred to as *signal integrity*).

From the point of view of the electronic designer, however, the interest lies describing the phenomenon in terms of voltage and current at the connection points of each circuit element, rather than in terms of electromagnetic fields at each point in space. Therefore, there is a great reduction in complexity. Furthermore, electronic semiconductor devices and components are modeled in

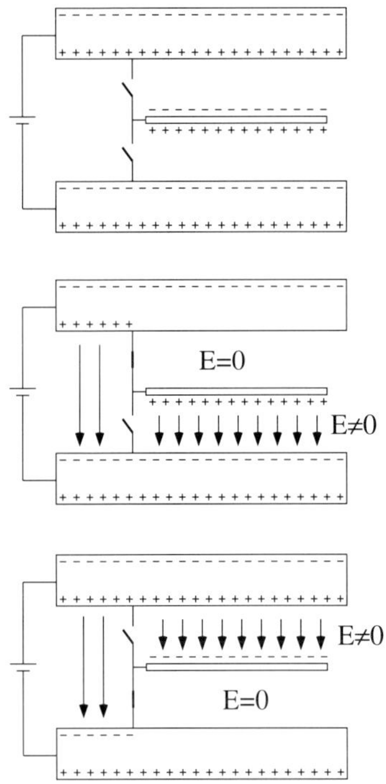

Figure 2.1. Ideal physical structure for describing a digital switching phenomenon.

terms of voltage-current responses and many methods for designing and analyzing electrical circuits have been developed and implemented in computer programs.

In this chapter, a brief justification of electrical models for interconnections will be presented, based on electromagnetic theory. This will enable us to understand the limitations and range of validity of such models, which will subsequently be used throughout this book.

1.1 TEM waves

Considering ideal conductors (null resistivity) and dielectrics (null conductivity), neither the electrical nor the magnetic fields will have components in the direction of the propagation of the wave. These waves are therefore called TEM, for Transverse ElectroMagnetic waves [1].

Taking as z axis the direction of propagation (figure 2.2) and following Maxwell's laws:

$$\begin{aligned} E_z &= 0 \\ H_z &= 0 \end{aligned}$$

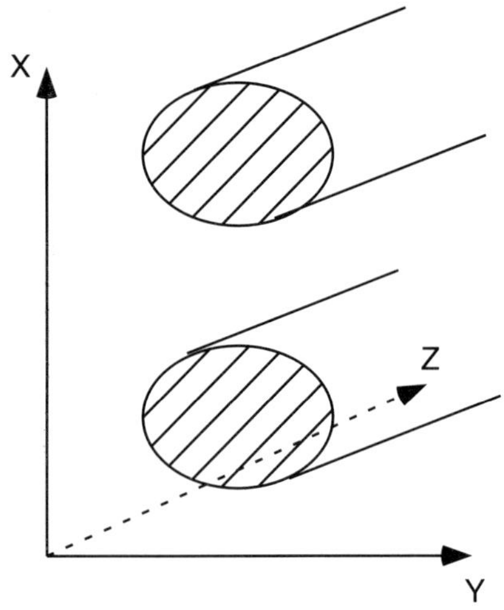

Figure 2.2. Physical structure of a waveguide.

$$\begin{aligned} \mathbf{E} &= E_x\hat{\mathbf{x}} + E_y\hat{\mathbf{y}} = \mathbf{E}_\mathrm{t} \\ \mathbf{H} &= H_x\hat{\mathbf{x}} + H_y\hat{\mathbf{y}} = \mathbf{H}_\mathrm{t} \end{aligned}$$

being $\mathbf{E}_\mathrm{t}$ and $\mathbf{H}_\mathrm{t}$ the transverse electric and magnetic fields respectively, and $\hat{\mathbf{x}}$ and $\hat{\mathbf{y}}$ the unit vectors in the x- and y-direction respectively.

The wave equation for the transverse electric and magnetic fields can be derived from the other components of vector equations 2.1 and 2.2 :

$$\frac{\partial^2 \mathbf{E}_\mathrm{t}}{\partial z^2} - \mu\epsilon \frac{\partial^2 \mathbf{E}_\mathrm{t}}{\partial t^2} = 0 \tag{2.3}$$

$$\frac{\partial^2 \mathbf{H}_\mathrm{t}}{\partial z^2} - \mu\epsilon \frac{\partial^2 \mathbf{H}_\mathrm{t}}{\partial t^2} = 0 \tag{2.4}$$

The solutions of the above equations are wave functions propagating in the z direction, with a velocity

$$v_p = \frac{1}{\sqrt{\mu\epsilon}} \tag{2.5}$$

1.1.1 Electrical model for TEM waves

In order to put the Maxwell's equations in more compact form, a transverse curl operator of a vector field with only $\hat{\mathbf{x}}$ and $\hat{\mathbf{y}}$ components, $\nabla_\mathrm{t} \times \mathbf{A}_\mathrm{t}$, is defined as:

$$\nabla_\mathrm{t} \times \mathbf{A}_\mathrm{t} = \hat{\mathbf{z}} \left(\frac{\partial A_y}{\partial x} - \frac{\partial A_x}{\partial y} \right) \tag{2.6}$$

From the $\hat{\mathbf{z}}$ component in equation 2.1, the following expressions are obtained for the electric field:

$$\nabla_{\mathbf{t}} \times \mathbf{E_t} = 0 \quad (2.7)$$

$$\frac{\partial E_x}{\partial z} = -\mu \frac{\partial H_y}{\partial t} \quad (2.8)$$

$$\frac{\partial E_y}{\partial z} = \mu \frac{\partial H_x}{\partial t} \quad (2.9)$$

Equation 2.7, means there is an equivalent to the electrostatic potential, $\Phi(x, y)$, in the plane orthogonal to the direction of propagation of the wave for each value of z. Correspondingly, a definition exists for the potential difference for each z:

$$V(z) = -\int_1^2 \mathbf{E_t} \cdot \mathbf{dl} \quad (2.10)$$

where the path of integration is arbitrary between conductors 1 and 2, as in the electrostatic case. This result justifies the calculation of the electrostatic field of the transverse structure for the analysis of uniform transmission lines (in the z-direction).

From equation 2.10, with 2.8 and 2.9:

$$\frac{\partial V}{\partial z} = -\int_1^2 \frac{\partial E_x}{\partial z} dx + \frac{\partial E_y}{\partial z} dy = -\int_1^2 -\frac{\partial B_y}{\partial t} dx + \frac{\partial B_x}{\partial t} dy =$$
$$-\frac{\partial}{\partial t} \int_1^2 -B_y dx + B_x dy = -\frac{\partial}{\partial t} \int_1^2 \mathbf{B} \cdot \mathbf{d\sigma} \quad (2.11)$$

where, vector $\mathbf{d\sigma} = \mathbf{dl} \times \hat{\mathbf{z}}$ is a differential vector orthogonal to the differential line vector $\mathbf{dl}$. It can then be interpreted as a surface-per-unit-length vector, and the last integral, as the surface integral of $\mathbf{B} = \mu \mathbf{H}$ per unit length in the direction of propagation. Now, using the definition of inductance:

$$L = \frac{1}{I} \int \mathbf{B} \cdot \mathbf{dS}$$

Equation 2.11 can be written as the first Telegrapher's equation:

$$\frac{\partial V}{\partial z} = -L \frac{\partial I}{\partial t} \quad (2.12)$$

where now L represents an inductance per unit length, and I the current in any one of the conductors in the direction of propagation.

Similarly, current in one of the conductors can be obtained from the usual definition of magnetostatics, assuming that current density is only non-zero inside the conductors and it is only in the direction of propagation ($\hat{\mathbf{z}}$):

$$\nabla_{\mathbf{t}} \times \mathbf{H_t} = J_z \hat{\mathbf{z}} \tag{2.13}$$

$$\frac{\partial H_x}{\partial z} = \epsilon \frac{\partial E_y}{\partial t} \tag{2.14}$$

$$\frac{\partial H_y}{\partial z} = -\epsilon \frac{\partial E_x}{\partial t} \tag{2.15}$$

Therefore

$$I = \oint \mathbf{H_t} \cdot \mathbf{dl} \tag{2.16}$$

and

$$\begin{aligned} \frac{\partial I}{\partial z} &= \oint \frac{\partial D_y}{\partial t} dx - \frac{\partial D_x}{\partial t} dy = \\ &-\frac{\partial}{\partial t} \oint -D_y dx + D_x dy = -\frac{\partial}{\partial t} \oint \mathbf{D} \cdot \mathbf{d}\sigma \end{aligned} \tag{2.17}$$

being $\mathbf{D} = \epsilon \mathbf{E}$ the displacement vector.

In this case, the path of integration is arbitrary and closed around any of the two conductors. As before, the last integral may be interpreted as a surface integral per unit length, and with Gauss's law and the definition of capacitance, the second Telegrapher's equation is obtained:

$$\frac{\partial I}{\partial z} = -C \frac{\partial V}{\partial t} \tag{2.18}$$

being C a capacitance per unit length in the direction of propagation.

It is easy to see that the two telegrapher's equations represent wave equations for voltage and current with a propagation constant given by γ. With a phasor formulation ($V = V(z)\exp(j\omega t)$, $I = I(z)\exp(j\omega t)$):

$$\frac{d^2 V(z)}{dz^2} = \gamma^2 V(z) \tag{2.19}$$

$$\frac{d^2 I(z)}{dz^2} = \gamma^2 I(z) \tag{2.20}$$

with $\gamma^2 = -LC\omega^2$. The solution is then

$$V(z) = V_o^+ \exp(-j\gamma z) + V_o^- \exp(+j\gamma z) \tag{2.21}$$

$$I(z) = I_o^+ \exp(-j\gamma z) + I_o^- \exp(+j\gamma z) \tag{2.22}$$

Using either equation 2.12 or 2.18, it is possible to find the relation between current and voltage amplitudes:

$$I_o^+ = \frac{V_o^+}{\sqrt{L/C}} \tag{2.23}$$

$$I_o^- = -\frac{V_o^-}{\sqrt{L/C}} \tag{2.24}$$

The relation between voltage and current amplitudes is defined as the *characteristic impedance*:

$$Z_o = \sqrt{\frac{L}{C}} \tag{2.25}$$

On the other hand, γ gives the velocity of propagation:

$$v_p = \frac{\omega}{\gamma} = \frac{1}{\sqrt{LC}} \tag{2.26}$$

Both magnitudes, Z_o and v_p are real and independent of frequency, which means there are no losses in propagation and the waves propagate undistorted. This case is therefore called the lossless or ideal transmission line.

The assumptions made to derive the telegrapher's equations from field equations are:

1 There are no field components in the direction of propagation. This is the case when there are no losses in the conductors nor in the dielectric.

2 Current density is exclusively in the direction of propagation, and zero outside the conductors.

1.2 Transmission lines with small losses (quasi-TEM model)

Real conductors and dielectrics will show losses due to finite conductivity and resistivity respectively. In this case, the conditions enumerated in the previous section no longer hold. However, if losses are small, it is still possible to find an approximate interpretation in terms of quasi-static voltage and current in the orthogonal plane, and, therefore, an electrical model [2], [3]. As it is obtained from a perturbation of the solution for the TEM model, it is called the quasi-TEM model.

$$\frac{\partial V}{\partial z} = -RI - L\frac{\partial I}{\partial t} \tag{2.27}$$

$$\frac{\partial I}{\partial z} = -GV - C\frac{\partial V}{\partial t} \tag{2.28}$$

where now R and G are series resistance and parallel conductance per unit length respectively, and they are obtained from a static analysis of the cross-section. In principle, therefore, their value is independent of frequency. We will see later that frequency-dependent parameters are considered for special cases.

The assumptions made to arrive at these equations are:

1 Current on the conductors flows only in the direction of propagation.

2 There is only a small perturbation of the fields with respect to the TEM solution.

Under these conditions, a formulation in terms of voltages and currents defined in the transverse plane of the structure can be found. The solution of the above equations for V and I is an attenuated wave propagating in the z-direction. The propagation constant is

$$\gamma = \sqrt{(R + j\omega L)(G + j\omega C)} \tag{2.29}$$

which is complex, meaning that there is an attenuation constant, α:

$$V(z) = V_o \exp(-\alpha z) \exp(\pm j\beta z) \tag{2.30}$$
$$I(z) = I_o \exp(-\alpha z) \exp(\pm j\beta z) \tag{2.31}$$

Now, the velocity of propagation, obtained from β, is dependent on frequency. There will, therefore, be some dispersion in the different frequency components of the signal, which means the signal will be distorted besides being attenuated.

The expression for characteristic impedance is:

$$Z_o = \sqrt{\frac{R + j\omega L}{G + j\omega C}} \tag{2.32}$$

This latter expression assumes implicitly $R \ll \omega L$ and $G \ll \omega C$ (small losses).

1.3 Simplifications of the *RLGC* model

The "complete" *RLGC* electrical model derived from physical assumptions can, in some circumstances, be simplified by neglecting some of the electrical parameters, In this way, the response given by the simplified model within a certain frequency range of interest is very similar to that of the *RLGC* model, at a much smaller computational cost. For the purpose of this book, we will classify electrical models as non-inductive or inductive.

1.3.1 Non-inductive models

At DC (zero frequency) and low frequency, given that the line termination is capacitive and the line is electrically isolated, no DC current flows through the interconnection. Thus, the simplified model consists of neglecting R, L and G, giving rise to a single (lumped) capacitive line model. At higher frequencies, as AC currents through the capacitors increase, the effect of line resistance starts to be significant, and, therefore, the RC model should be considered. For the lower range of frequencies, a single resistor and capacitor is accurate enough (lumped RC model), but as frequency increases, more RC stages may be necessary, and at the limit, the distributed RC model is made up of an infinite number of RC stages, with equations corresponding to the *RLGC* model with null L and G:

$$\frac{\partial V}{\partial z} = -RI \tag{2.33}$$

$$\frac{\partial I}{\partial z} = -C\frac{\partial V}{\partial t} \tag{2.34}$$

This equation corresponds to a diffusion equation and no longer to wave propagation. It is therefore physically incorrect and should be regarded as a mathematical simplification for the sake of simplicity.

In general, all solutions of the non-inductive models have in common that they do not describe oscillations with a monotonic excitation. These models are therefore inadequate under ringing (oscillations due to successive reflections) conditions, as will be discussed later.

1.3.2 Inductive models

For still higher frequencies, line impedance is not purely resistive, but has a non-negligible AC component, and a line model with inductance is necessary. Conductance can be neglected up to a certain frequency limit, when the complete, more physical *RLGC* model is necessary. This will be valid while the physical quasi-TEM approximation is valid.

In principle it is possible to consider a lumped RLC model, but, given the frequency range involved, it usually makes more sense to consider a distributed RLC model, with equations:

$$\frac{\partial V}{\partial z} = -RI - L\frac{\partial I}{\partial t} \tag{2.35}$$

$$\frac{\partial I}{\partial z} = -C\frac{\partial V}{\partial t} \tag{2.36}$$

These equations are wave equations for V and I, and they are therefore more accurate.

Figure 2.3 shows the validity range qualitatively in terms of frequency of the models following the above classification and indicating the complexity

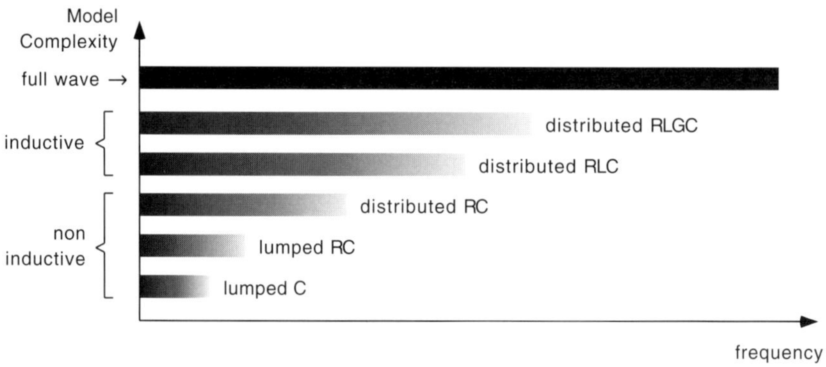

Figure 2.3. Validity range in terms of frequency of different line models: non-inductive, inductive and full wave (non electrical, valid for any frequency).

(in computational cost) of each type of model. The maximum frequency of validity for each model is not exact, because it also depends on the accuracy required. For example, a digital simulation in the early stages of physical design will probably need only a rough estimate of delay and therefore a capacitive model will be enough.

It should be noted that the frequency for validity for each model refers here to the frequency spectrum of the signal propagating inside the line, not at the driver input. This is an important distinction, because parameters like line resistance, and driver resistance can attenuate high-frequency components of input signals, and the propagating waves will show a lower frequency content, leading to the possibility of using a simpler model. This issue will be examined in a later section.

1.4 Validity of the quasi-TEM assumption

As explained before, the quasi-TEM model was derived using several assumptions about the propagating fields. In reality, both conductors forming a transmission line will have finite conductivity. This means the current in the conductors will generate a longitudinal electric field due to Ohm's law. In addition, the skin effect will concentrate the conductor longitudinal current at the periphery of the conductor, causing an increase in resistance and a decrease in inductance for high frequencies. All these effects, if significant, may lead to deviations from pure TEM wave propagation and therefore, make the electrical models derived from it invalid.

Firstly, it is important to try to determine the limits of the electrical model from physical dimensions. Several papers have addressed this problem [4], [5]. In addition, Brews [6] presented a way of deriving an electrical model even for high loss, non-TEM waveguides.

Mesa et al. [4], [7] presented a dimensional analysis to assert the validity of the quasi-TEM analysis for a waveguide structure with substrate losses. This is a practical case, since in all integrated circuits the substrate is a semiconductor with moderate conductivity, which, at least in principle, can invalidate quasi-TEM assumptions. In the case in which the losses are very significant, a full-wave analysis is necessary for analyzing signal transmission.

The analysis is based on the condition that the longitudinal fields E_z and H_z are much smaller than their transverse components, but on a spatial average rather than on a point-to-point basis.

$$\langle |E_z| \rangle \ll \langle \|\mathbf{E_t}\| \rangle \quad \langle |H_z| \rangle \ll \langle \|\mathbf{H_t}\| \rangle \tag{2.37}$$

where $\langle \bullet \rangle$ means spatial average, $| \bullet |$ means modulus, and $\| \bullet \|$ means vector norm.

The analysis in [7] is for lines embedded in general media which can present dielectric and magnetic anisotropy, and generally both magnetic permeability and dielectric permittivity will be complex ($\hat{\mu}$ and $\hat{\epsilon}$ respectively). The condition obtained is related to characteristic transverse dimensions of the line structure d, to μ and ϵ, and to signal frequency, ω as:

$$d \ll \frac{1}{\omega\sqrt{\langle |\hat{\mu}(\mathbf{r})| \rangle \langle |\hat{\epsilon}(\mathbf{r})| \rangle}} \tag{2.38}$$

where, in the above equation, the spatial averages for $\hat{\mu}$ and $\hat{\epsilon}$ are in the region where the fields are non-zero (or at least, sufficiently important).

For the usual simpler cases of CMOS integrated circuit technology, in which magnetic permeability is practically the same for all materials, only the permittivity varies with position, $\mathbf{r}$. For SiO_2 or air, the permittivity is real, while for silicon substrate, it will be complex because of its conductivity (0.01 S/mm to 0.02 S/mm for current technologies):

$$\hat{\epsilon} = \epsilon + \frac{\sigma}{j\omega} \tag{2.39}$$

Hence, for the cases of silicon integrated circuits:

$$d \ll \frac{1}{\omega\sqrt{\mu_0 \langle \epsilon(\mathbf{r}) \rangle \sqrt{1 + \frac{\langle \sigma(\mathbf{r}) \rangle^2}{\omega^2 \langle \epsilon(\mathbf{r}) \rangle^2}}}} \tag{2.40}$$

Now, let us make some rough but reasonable estimates of the validity of the quasi-TEM model for lines in silicon ICs.

Let us take some extreme cases. The first case to consider will be one in which the fields are far away from the substrate. In this case, conductivity is essentially zero in the region of interest, and therefore, so it is its spatial average. In general, this situation can be expressed as:

$$\langle \sigma(\mathbf{r}) \rangle \ll \omega \langle \epsilon(\mathbf{r}) \rangle) \tag{2.41}$$

Under this condition, the range of validity for the quasi-TEM model is:

$$d \ll \frac{1}{\omega \sqrt{\mu_0 \langle \epsilon(\mathbf{r}) \rangle}} \tag{2.42}$$

The spatial average of permittivity will have a value between that of air $\epsilon_{air} \approx 8.9$ pF/m and the value for SiO_2, $\epsilon_{SiO2} \approx 35.4$ pF/m. The most restrictive situation is that of a spatial average close to 35.4 pF/m. If we take 100 μm as a typical value for the dimension of a transmission line structure cross-section, the validity of quasi-TEM model holds for signal frequencies such that:

$$\omega \ll 1.5 \cdot 10^{12}\,\text{rad/s} = 238\,\text{GHz}$$

Such a frequency limit is beyond significant frequencies of digital signals in integrated circuits today and for the near future, so the quasi-TEM model is clearly applicable in these cases.

Now, let us consider the general case in which the fields extend to the semiconductor substrate. In this case, the spatial average should consider the region of the silicon substrate as well as the SiO_2 and, possibly, air. Hence, the spatial average conductivity will not be zero, but a value of between, say, 20 S/m and 0 S/m for current CMOS technologies. The spatial average permittivity will also increase, because silicon presents a higher real permittivity. In order to obtain a rough estimate for the range of validity, let us model the spatial average permittivity and conductivity as a function of a parameter: effective relative permittivity, ϵ_{reff}. We will assume that when ϵ_{reff} is close to that of SiO_2, then the average conductivity will approach zero. When ϵ_{reff} is close to that of Si substrate, the conductivity will reach the value for silicon, which we take as 20 S/m in this example. Although simplified, this is an intuitive behavior. The results are shown in figure 2.4, where *dmax* represents the maximum cross-section dimension such that the quasi-TEM approximation is still valid, in function of signal frequency propagating in the interconnection structure. Here, two regions can be seen: for low frequencies and finite conductivity the dependence of d is like $1/\sqrt{\omega}$. For high frequencies, it goes like $1/\omega$, as the case when conductivity is zero.

In any case, it is seen that the frequency limit for a distance of $100\,\mu\text{m}$ is of the order of $100\,\text{GHz}$, which validates the quasi-TEM model with $RLGC$ parameters in the majority of cases for current CMOS integrated circuits.

2. Lossless transmission line model

Now we have introduced the physical foundations for the electrical models, we will concentrate on examining the response of these electrical models to

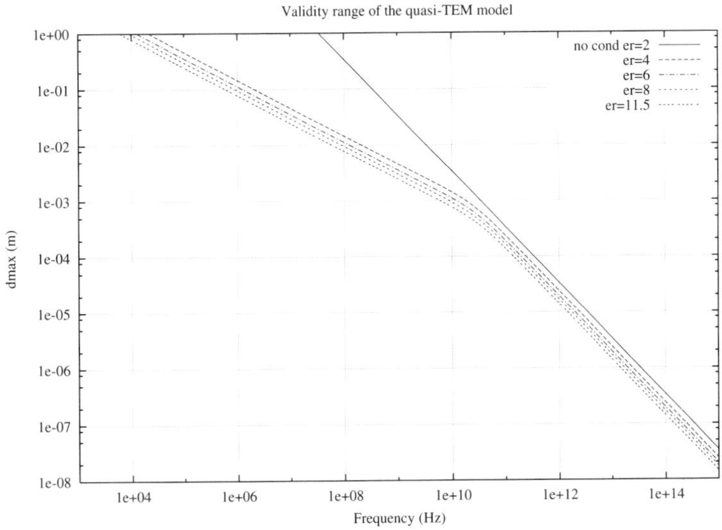

Figure 2.4. Graph of maximum dimension such that the quasi-TEM model is valid, following expression 2.40. The effective relative permittivity is varied. Also shown the graph for zero average conductivity.

common situations found in integrated circuits. The objective of the following sections is to try to find a set of rules for selecting a model which is both accurate and simple for simulating the conditions found (type of driver, load, line length and frequency characteristics of the signal to be transmitted).

One effect of signal transmission in interconnections is the appearance of oscillations (ringing). This kind of voltage oscillation may induce transient errors, or in general, a distortion of the original transmitted waveform. They are physically related to reflections of the electromagnetic wave, which translate in the electrical model into reflections of voltage and current waves.

Firstly, the resistive driver model will be considered for an analysis of ringing. Then, the effect of signal rise time in resistive drivers will be considered and, finally, a more realistic CMOS driver model will be analyzed.

2.1 Resistive driver model

Whenever an incident voltage wave encounters a discontinuity in impedance, such as a change in line parameters or the impedance of the load of the line (figure 2.5), part of the wave is reflected and part of it is transmitted. From the figure it can be written:

$$V_t = V_i + V_r \tag{2.43}$$
$$I_t = I_i - I_r \tag{2.44}$$

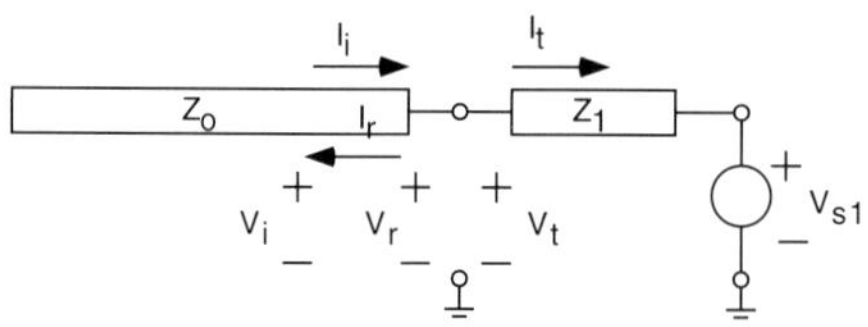

Figure 2.5. Impedance discontinuity in a line

Applying the definition of characteristic impedance, we can find the expression of the transmitted and reflected voltages for a general case where the incident voltage in the transmission line (impedance Z_o) arrives at an end (impedance Z_1) where there is an arbitrary fixed voltage source, V_{s1}. From the equation for currents 2.44:

$$\begin{aligned}\frac{V_i}{Z_o} &= \frac{V_t - V_{s1}}{Z_1} + \frac{V_r}{Z_o} = \\ &\frac{V_i}{Z_1} + V_r\left(\frac{1}{Z_1} + \frac{1}{Z_0}\right) - \frac{V_{s1}}{Z_1}\end{aligned}$$

that is,

$$V_r = V_i\frac{Z_1 - Z_o}{Z_1 + Z_o} + V_{s1}\frac{Z_o}{Z_o + Z_1} \tag{2.45}$$

$$V_t = V_i\frac{2Z_1}{Z_1 + Z_o} + V_{s1}\frac{Z_o}{Z_o + Z_1} \tag{2.46}$$

The dependence of the reflected and transmitted voltages on the incident voltage is determined respectively by the reflection (ρ) and transmission (τ) coefficients:

$$\rho = \frac{Z_1 - Z_o}{Z_1 + Z_o} \tag{2.47}$$

$$\tau = \frac{2Z_1}{Z_l + Z_o} = 1 - \rho \tag{2.48}$$

The final waveform at the end of the line is given by the sum of the voltages due to successive reflections. In order to illustrate the calculation of the successive voltages at the driver and load ends of the line, we consider the example of a line with characteristic impedance $Z_o = 100\,\Omega$ and generic time of propagation t_p, with a driver impedance $Z_d = 75\,\Omega$, terminated in open circuit ($Z_l = \infty$). Note that $Z_d < Z_o$. In these conditions, the response to an ideal step waveform of amplitude $V_e = 1$ V (figure 2.6) will be given for subsequent time steps as:

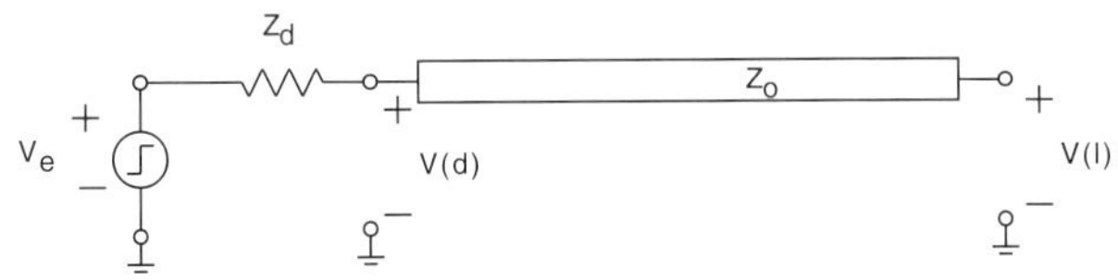

Figure 2.6. A line terminated in open circuit

1 At $t = 0$, incident and reflected waves are zero, so from equation 2.45 the voltage at the driver and load ends are:

$$\begin{aligned} V_0(d) &= V_e \frac{Z_o}{Z_o + Z_d} = 0.571\,\mathrm{V} \\ V_0(l) &= 0 \end{aligned}$$

2 At $t = t_p$, incident voltage amplitude at the load end of the line is $V_0(d)$, and transmitted and reflected amplitudes are given by equations 2.46 and 2.45 respectively. The voltage at the driver and load ends is:

$$\begin{aligned} V_1(d) &= V_0(d) = 0.571\,\mathrm{V} \\ V_1(l) &= V_0(d)\frac{2Z_l}{Z_l + Z_o} = 1.142\,\mathrm{V} \\ V_{1r} &= V_0(d)\frac{Z_l - Z_o}{Z_l + Z_o} = 0.571\,\mathrm{V} \end{aligned}$$

3 At $t = 2t_p$, incident amplitude at the driver end of the line is V_{1r} (reflected wave from load end) and transmitted and reflected amplitudes are again given by equations 2.46 and 2.45 respectively. The voltage at the driver and load ends is:

$$\begin{aligned} V_2(d) &= V_{1r}\frac{2Z_d}{Z_d + Z_o} + V_e\frac{Z_o}{Z_o + Z_d} = 1.060\,\mathrm{V} \\ V_2(l) &= V_1(l) = 1.142\,\mathrm{V} \\ V_{2r} &= V_{1r}\frac{Z_d - Z_o}{Z_d + Z_o} + V_e\frac{Z_o}{Z_o + Z_1} = 0.489\,\mathrm{V} \end{aligned}$$

4 At $t = 3t_p$, incident amplitude at the load end of the line is V_{2r} and the voltage at the driver and load ends are:

$$\begin{aligned} V_3(d) &= V_2(d) = 1.060\,\mathrm{V} \\ V_3(l) &= V_{2r}\frac{2Z_l}{Z_l + Z_o} = 0.978\,\mathrm{V} \\ V_{3r} &= V_{2r}\frac{Z_l - Z_o}{Z_l + Z_o} = 0.489\,\mathrm{V} \end{aligned}$$

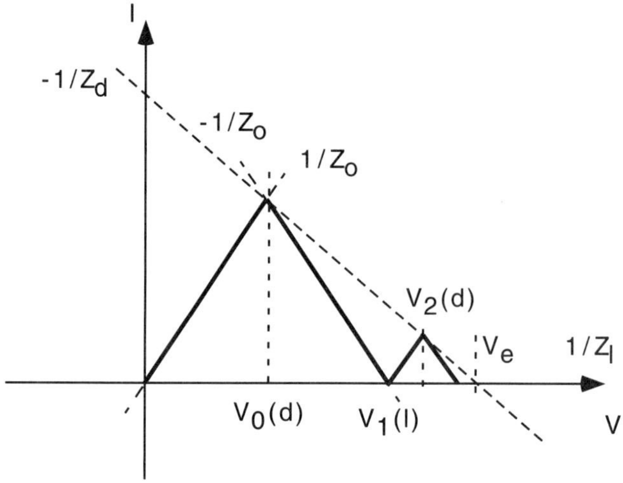

Figure 2.7. Graph method for calculating reflections in the line. In this case, the load impedance is zero (open circuit termination).

5 At $t = 4t_p$, incident amplitude at the driver end of the line is V_{3r} and the voltage at the driver and load ends are:

$$\begin{aligned} V_4(d) &= V_{3r}\frac{2Z_d}{Z_d + Z_o} + V_e\frac{Z_o}{Z_o + Z_d} = 0.990\,\text{V} \\ V_4(l) &= V_3(l) = 0.978\,\text{V} \\ V_{4r} &= V_{3r}\frac{Z_d - Z_o}{Z_d + Z_o} + V_e\frac{Z_o}{Z_o + Z_1} = 0.501\,\text{V} \end{aligned}$$

As shown, the subsequent voltage amplitudes at driver and load ends oscillate and converge to a steady value after a number of iterations which depend on the reflection coefficients at the driver and load ends, ρ_d and ρ_l given by:

$$\rho_d = \frac{Z_d - Z_o}{Z_d + Z_o} \tag{2.49}$$

$$\rho_l = \frac{Z_l - Z_o}{Z_l + Z_o} \tag{2.50}$$

However, the response is not always an oscillation of voltages (ringing). In order to see the condition for ringing intuitively, it is possible to use a graph method for determining the waveform after reflections [8], that we will call here Method of Characteristics. In figure 2.7 a case is presented for illustration considering linear (resistive) impedances of driver and load.

Firstly, a straight line is drawn with slope $-\frac{1}{Z_d}$, being Z_d the driver impedance, and such that this line cuts the voltage axis at a value equal to the step

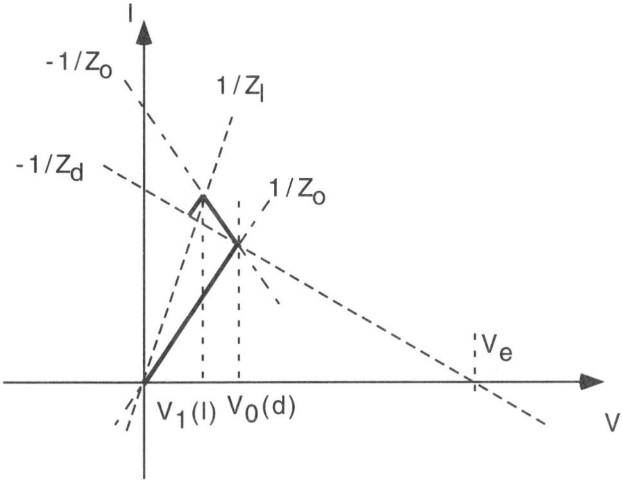

Figure 2.8. Diagram for a line with $Z_l < Z_d$, and $Z_l < Z_o < Z_d$. The response is a voltage oscillation around the steady state value.

amplitude, V_e. The initial voltage and current at the driver end of the line is given by the intersection of this straight line and the one corresponding to the characteristic impedance, with slope $\frac{1}{Z_o}$. It is easy to see that this first point corresponds to expression $V_0(d)$. Drawing another straight line from this point, with slope $\frac{-1}{Z_o}$, the values for voltage and current at the load end of the line ($t = t_p$) are obtained from its intersection with a second straight line with slope $\frac{1}{Z_L}$, corresponding to the load impedance (which in the case of figure 2.7 is a horizontal line because the line is terminated in open circuit). The voltage again coincides with the analytical method voltage, $V_1(l)$. Repeating this process successively, the different values for voltage and current are obtained graphically.

This method enables the impedance values for line, driver and load for which there will be oscillations (ringing) to be predicted easily. It is important to identify this situation, because it may give rise to problems due to the extra time the signal takes to become stable and may cause false commutation events in digital circuits.

If the driver and load have linear characteristics, the condition for the appearance of ringing is (figures 2.8 and 2.10):

$$\text{If } Z_l > Z_d: \quad \left.\begin{array}{r} \frac{1}{Z_o} \leq \frac{1}{Z_d} \\ \text{and } \frac{1}{Z_o} \geq \frac{1}{Z_L} \end{array}\right\} \quad Z_d \leq Z_o \leq Z_l \tag{2.51}$$

$$\text{If } Z_l < Z_d: \quad \left.\begin{array}{r} \frac{1}{Z_o} \geq \frac{1}{Z_d} \\ \text{and } \frac{1}{Z_o} \leq \frac{1}{Z_l} \end{array}\right\} \quad Z_d \geq Z_o \geq Z_l \tag{2.52}$$

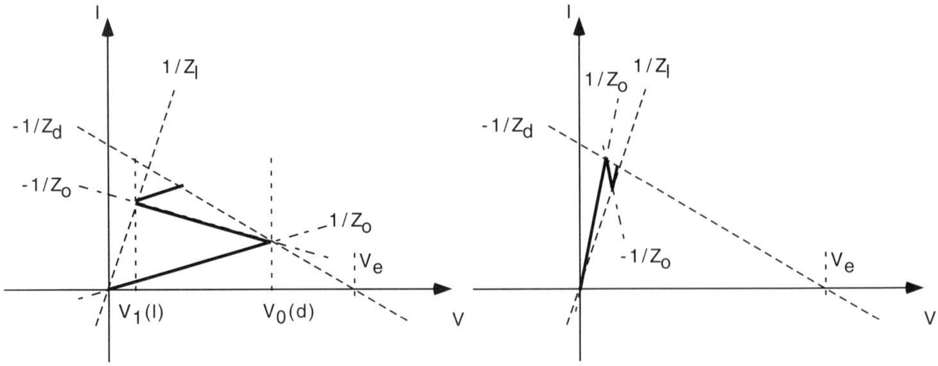

Figure 2.9. Diagram for a line with $Z_l < Z_d$, with $Z_o > Z_d$ (left), and $Z_o < Z_l$ (right). The response tends asymptotically towards the steady value without any oscillations, although an initial overshoot at the driver end is observed in the first case.

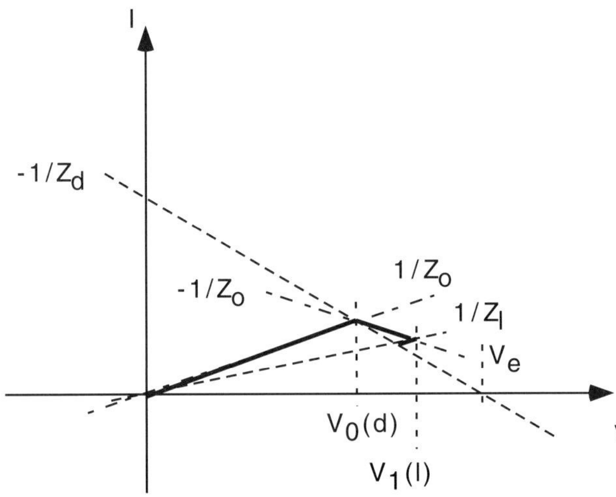

Figure 2.10. Diagram for a line with $Z_l > Z_d$, and $Z_l > Z_o > Z_d$. The response is a voltage oscillation around the steady state value.

Therefore, ringing will appear if the value of Z_o is in between of driver and load impedances, Z_d, and Z_l. In other cases (illustrated by figures 2.9 and 2.11), ringing is not produced.

2.2 Effect of signal rise time

The amplitude of the oscillations in the case of ringing is maximum for ideal step transitions (zero rise time). For rise times much greater than that, oscillations are effectively suppressed and a simpler non-inductive line model would suffice to describe the response of the interconnection accurately.

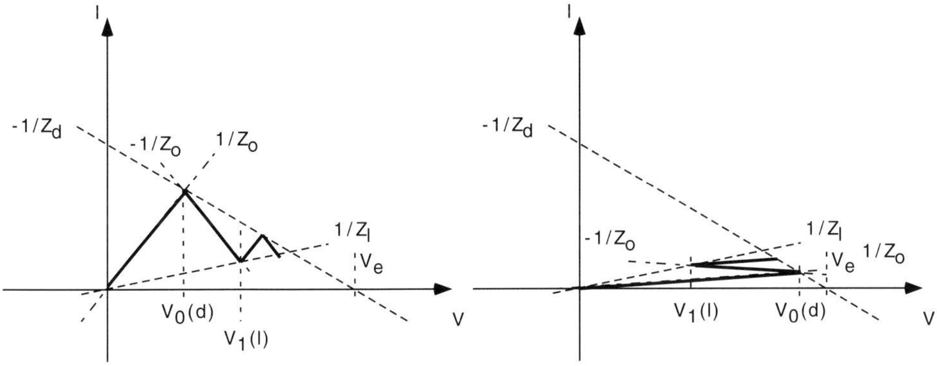

Figure 2.11. Diagram for a line with $Z_l > Z_d$, with $Z_o < Z_d$ (left), and $Z_o > Z_l$ (right). The response tends asymptotically to the steady value without any oscillations.

It is possible to find an expression to find the rise time value that will cause total suppression of voltage oscillations, by using the method of characteristics shown above. For the sake of simplicity, let us consider a resistive driver. Let us assume that the input voltage changes linearly from 0 to V_e over a time t_r. Using the graph method of characteristics, this means that the driver characteristic is displaced during time t_r from the origin to its final position (figure 2.12). Depending on the number of times the wave travels back and forth in the interconnection before the input voltage reaches its steady value, ringing can be eliminated. Knowing that in the case of ringing the maximum peak voltage is during the first reflection (from $t = t_p$ to $t = 3t_p$), we can find the condition for non-ringing by imposing that this maximum voltage must be smaller than V_e. Unlike the ideal case of zero rise time, the load end voltage during a period t_p is not constant, but changes, as does the driver end voltage. We are therefore interested in load voltage at time $t - 3t_p$, and the condition for a non-ringing waveform is:

$$V(x = l, t = 3t_p) \leq V_e \tag{2.53}$$

$$V(x = l, t) = V(x = 0, t - t_p)\frac{2Z_l}{Z_l + Z_o} \tag{2.54}$$

$$V(x = 0, t) = V_s(t)\frac{Z_o}{Z_o + Z_d} \tag{2.55}$$

Thus, assuming a ramp input:

$$V_s(t) = \begin{cases} V_e\frac{t}{t_r} & \text{for } 0 < t < t_r \\ V_e & \text{for } t \geq t_r \end{cases} \tag{2.56}$$

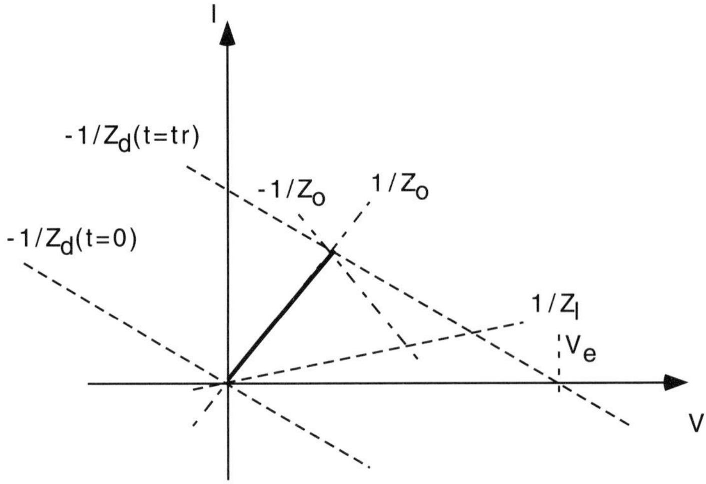

Figure 2.12. Displacement of the driver characteristic during a finite rise time input transition.

$$V(x = l, t = 3t_p) = \begin{cases} V_e \frac{2t_p}{t_r} \frac{2Z_l}{Z_l + Z_o} \frac{Z_o}{Z_o + Z_d} & \text{if } t_r > 2t_p \\ V_e \frac{2Z_l}{Z_l + Z_o} \frac{Z_o}{Z_o + Z_d} & \text{if } t_r \leq 2t_p \end{cases} \tag{2.57}$$

For the case $Z_l = \infty$ (open circuit), imposing condition 2.53 for non-ringing response gives:

$$Z_d \geq \begin{cases} Z_o \left(\frac{4t_p}{t_r} - 1 \right) & \text{if } t_r > 2t_p \\ Z_o & \text{if } t_r \leq 2t_p \end{cases} \tag{2.58}$$

Note that the non-ringing condition for $t_r \leq 2t_p$ is the same as the particular case of zero rise time and open circuit load. Also note that, for $t_r > 4t_p$, ringing is impossible for any value of Z_d (which is always positive).

2.3 Lossless lines with a CMOS driver and open circuit load

Another advantage of the graph method is that it enables easy consideration of drivers and loads with non-linear characteristics, that is, those whose impedance varies with voltage. As an important example, a CMOS inverter is considered as the driver in this section (figure 2.13).

In order to cause a transition 0–V_{DD} at the inverter output, V_{in} is assumed to change instantly from V_{DD} to 0, making it consequently the NMOS transistor cut-off. Following Sah model MOS transistor equations, due to the fact that $V_{in} = 0$, there will be two different zones of current behavior depending on the inverter's output voltage $V(d)$:

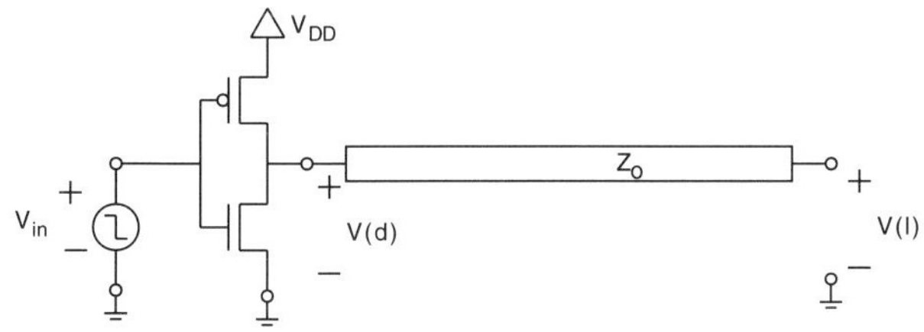

Figure 2.13. Schematic of a line with a CMOS inverter as a driver

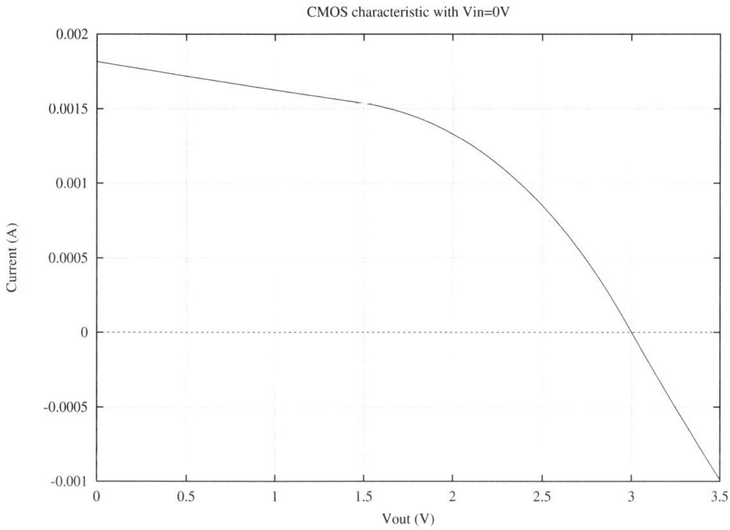

Figure 2.14. Output characteristic of a CMOS inverter for $V_{in} = 0$ V and $V_{DD} = 3$ V.

- Saturation region, $V(d) \leq |V_{tp}|$:

$$I = \frac{K_p}{2}\left(V_{DD} - |V_{tp}|\right)^2 \tag{2.59}$$

- Linear region, $V(d) \geq |V_{tp}|$:

$$I = K_p\left(V_{DD} - |V_{tp}| - \frac{V_{DD} - V(d)}{2}\right)\left(V_{DD} - V(d)\right) \tag{2.60}$$

These equations give rise to a non-linear characteristic represented in figure 2.14.

As schematically shown in figure 2.15, ringing will occur when:

$$V_0(d) \geq \frac{V_{DD}}{2} \tag{2.61}$$

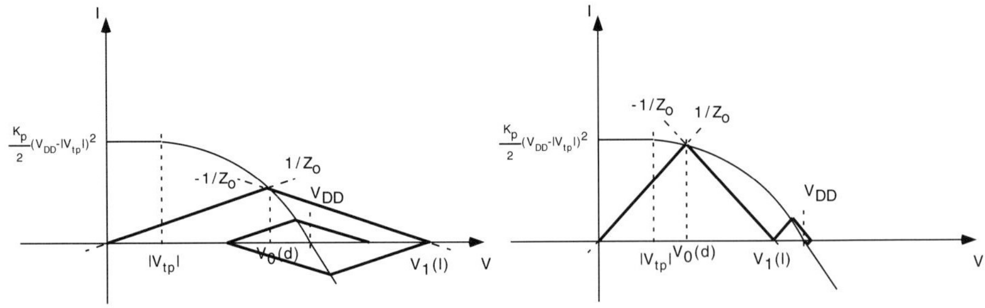

Figure 2.15. Method of Characteristics with CMOS nonlinear impedance. If $V_0(d) < V_{DD}/2$, only minor ringing is possible.

being $V_0(d)$ the first amplitude in the line. Actually, it is the same condition as for a linear driver when the load has infinite impedance, because in that case $V_0(d) = V_{DD}Z_d/(Z_d + Z_o)$, and condition 2.61 reduces to $Z_d \geq Z_o$.

This condition can be translated to a condition relating circuit parameters, as shown in the following paragraphs. We will distinguish two situations:

2.3.1 Case $V_{DD} \geq 2|V_{tp}|$

If $V_{DD} \geq 2|V_{tp}|$, $V_0(d)$ can only be greater than $V_{DD}/2$ in the linear region. The expression for $V_0(d)$ is obtained from:

$$\frac{V_0(d)}{Z_o} = K_p \left(V_{DD} - |V_{tp}| - \frac{V_{DD} - V_0(d)}{2} \right) (V_{DD} - V_0(d)) \qquad (2.62)$$

From this last expression, condition 2.61 translates to:

$$\frac{1}{Z_o K_p} \leq \frac{3}{4} V_{DD} - |V_{tp}| \qquad (2.63)$$

2.3.2 Case $V_{DD} \leq 2|V_{tp}|$

For such low values of V_{DD} with respect to $|V_{tp}|$, condition 2.61 corresponds to an intersection in the saturation region:

$$\frac{V_0(d)}{Z_o} = \frac{K_p}{2}(V_{DD} - |V_{tp}|)^2 \qquad (2.64)$$

And the condition for $1/Z_o K_p$:

$$\frac{1}{Z_o K_p} \leq \frac{(V_{DD} - |V_{tp}|)^2}{V_{DD}} \qquad (2.65)$$

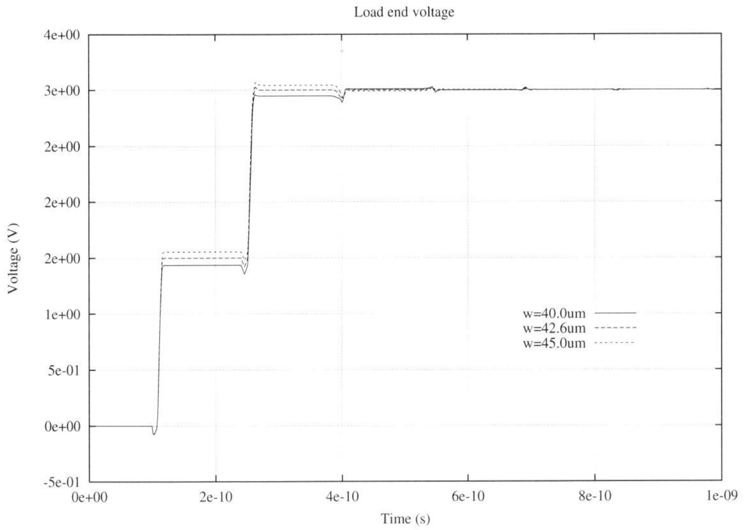

Figure 2.16. HSPICE simulation comparing the response to three different sizes of PMOS transistor ($L = 0.25\,\mu\text{m}$, $K'_p = 34\,\mu\text{A/V}^2$, $V_{tp} = -0.8\,\text{V}$, and $V_{DD} = 3\,\text{V}$). Z_o of the line is 119 Ω.

Figure 2.16 shows a simple demonstration simulation of an 119 Ω transmission line and a CMOS driver with PMOS having L=0.25 μm, K'_p=34 μA/V^2, $V_{tp} = -0.8\,\text{V}$, and $V_{DD} = 3\,\text{V}$. Equation 2.63 gives a value of W equal to 42.6 μm for series termination, and the figure shows that values of W different than this value cause over- or undershoots.

3. Lossy transmission line model

In the preceding sections, the response of lossless lines was analyzed to evaluate the importance of reflections on response to a voltage change. This lossless line model is a reasonable approximation for off-chip lines; at PCB level, and it is the usual model for the analysis of, for example, termination schemes [9].

However, for interconnections inside the chip, as technology scales down, line resistance cannot be ignored. In addition, the semiconductor substrate is not a perfect conductor. Therefore, a lossy model must be used to evaluate the response of the interconnections.

The effect of losses can be understood as a perturbation of the lossless case response. The effect will be to attenuate oscillations and to slow down the response. In this respect, then, the previous section's analysis is already useful: the lossless model conditions for ringing are a worst case when considering a lossy case. That is, when oscillations are not possible in the lossless model,

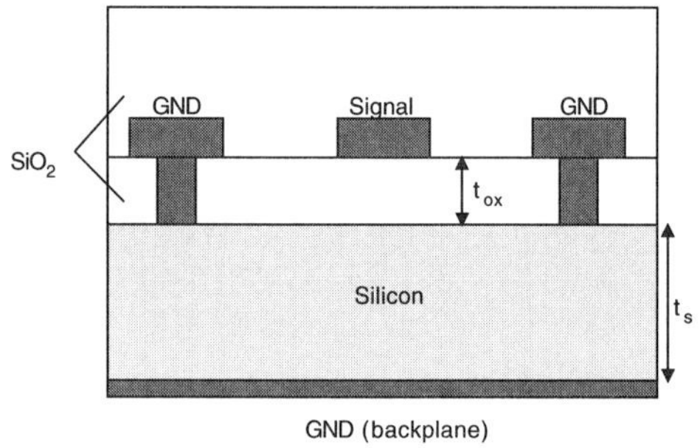

Figure 2.17. Vertical structure of a line over semiconductor substrate, with reference contacts on the top and bottom surfaces.

they will not be possible either with the lossy model. When ringing is significant in the lossless case, the lossy effect may be significant, depending on the magnitude of losses. Ho et al. [10] derived a limiting expression for RLC lines, comparing the value of $R \cdot l$ (total resistance of the line) with the value of the lossless characteristic impedance, $Z_o = \sqrt{L/C}$. They found that a total resistance of the line $R \cdot l$ greater than approximately $2.5Z_o$ would attenuate to negligible levels the possible ringing, and therefore the inductive effect of the line. This type of *a priori* analysis allows to select an interconnect model which is simpler than the complete transmission line and reduce the complexity of the simulation task.

In the following paragraphs, the effect of semiconductor substrate losses will be illustrated, giving some background theory, and some simulation examples.

3.1 Effect of semiconductor substrate

In integrated circuits, interconnections are made by metallic deposition over a silicon dioxide or other type of insulator between the interconnection and the semiconductor substrate (figure 2.17).

The semiconductor substrate is connected to the reference voltage through an electrical contact on top, or in some technologies, through back metallization covering the entire bottom surface of the substrate. In some chip configurations and depending on the packaging options, the backplane is not electrically connected to the reference voltage, but rather left floating.

The semiconductor substrate introduces several effects that affect interconnection modeling [11]. Physically, the capacitance of the line is a combination of the capacitance to the substrate in series with the capacitance of the semiconductor, C_{ox} and C_s respectively (figure 2.18), plus the capacitance to the parallel power/ground lines, C_p.

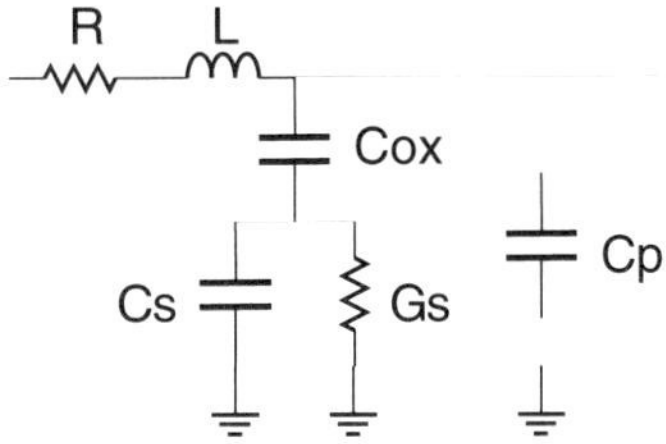

Figure 2.18. Electrical diagram of the parameters of a line over semiconductor substrate.

Also, the semiconductor has a non-zero conductivity that depends on doping, giving rise to a conductance represented by G_s.

All these components show an equivalent admittance to ground that can be expressed as:

$$Y(\omega) = \frac{\omega^2 C_{ox}^2 G_s}{G_s^2+\omega^2(C_{ox}+C_s)^2} + j\omega\left(C_p + \frac{C_{ox}G_s^2+\omega^2 C_{ox}C_s(C_{ox}+C_s)}{G_s^2+\omega^2(C_{ox}+C_s)^2}\right) \quad (2.66)$$

Therefore, this admittance can be identified as lossy *RLGC* transmission line admittance with frequency-dependent conductance and capacitance, $Y(\omega) = G(\omega) + j\omega C(\omega)$, with $G(\omega)$ and $C(\omega)$:

$$G(\omega) = \frac{\omega^2 C_{ox}^2 G_s}{G_s^2 + \omega^2(C_{ox} + C_s)^2} \quad (2.67)$$

$$C(\omega) = C_p + \frac{C_{ox}G_s^2 + \omega^2 C_{ox}C_s(C_{ox} + C_s)}{G_s^2 + \omega^2(C_{ox} + C_s)^2} \quad (2.68)$$

Depending on the frequency of the signal, the line has different operational regimes:

- Slow wave mode: for low frequencies, the equivalent line conductance $G(\omega) \approx 0$, and the equivalent capacitance $C(\omega) \approx C_{ox} + C_p$. This can be interpreted in the sense that the skin depth in the semiconductor substrate $\delta = \sqrt{2\rho_s/\omega\mu_0}$ is greater than substrate thickness t_s, and the substrate conductivity G_s acts like a short-circuit, bypassing capacitance C_s (see figure 2.18). Inductance depends on the distances of the line from ground lines, s, as well as to the backplane, $t_{ox} + t_s$.

- Skin depth mode: for mid-range frequencies, the return current path is inside the semiconductor; inductance is then determined by distance $t_{ox} + \delta$.

- Dielectric mode: for high frequencies ($\omega \gg \frac{G_s}{C_{ox}+C_s}$), the equivalent line conductance is $G(\omega) \approx \frac{C_{ox}^2 G_s}{(C_{ox}+C_s)^2}$, and the equivalent capacitance is $C(\omega) \approx C_p + \frac{C_{ox}C_s}{C_{ox}+C_s}$.

Firstly, we consider the case where no ground line is present nearby. In this case, $C_p = 0$, and a rough estimate of the other parameters involved can be obtained with the following expressions:

$$G_s \approx \frac{W \cdot L}{\rho_s t_s} \tag{2.69}$$

$$C_s \approx \frac{k_s \epsilon_0}{t_s} W \cdot L \tag{2.70}$$

$$C_{ox} \approx \frac{k_{ox} \epsilon_0}{t_{ox}} W \cdot L \tag{2.71}$$

being W the line width, L the line length, ρ_s substrate resistivity, k_s and k_{ox} the dielectric constants of substrate and SiO_2 respectively.

To determine the order of magnitude of the transition frequency for each mode, a numerical example is presented for a silicon wafer 500 μm thick. If silicon is doped p-type with $N_A = 2 \cdot 10^{16} \mathrm{cm}^{-3}$ its resistivity will be:

$$\rho_s \approx (q N_A \mu_p)^{-1} \approx 0.65\,\Omega \cdot \mathrm{cm}$$

The value of ω for which δ equals substrate thickness is:

$$\omega_{sw} = \frac{2\rho}{t_s^2 \mu_0} = 4.1 \cdot 10^{10} \mathrm{rd/s}$$

which corresponds to a signal frequency $\nu_{sw} = 6.6$ GHz. For lower frequencies, the skin depth δ is greater than the substrate thickness, so the return current path is through the backplane and not through the semiconductor substrate for the majority the digital signal spectrum.

For this configuration, a $10\,\mu\mathrm{m}$ wide line has the following values per unit length: $G_s \approx 3\,\mathrm{S/m}$, $C_s \approx 2.06\,\mathrm{pF/m}$, $C_{ox} \approx 343.98\,\mathrm{pF/m}$.

When there is a nearby ground line connected to substrate, two main differences are to be noted: substrate conductance G_s increases because the distance to nearest ground is reduced, and a parallel direct capacitance to ground appears, C_p. For conductance, again, a rough estimate can be made with the following expression:

$$G_s \approx \frac{t_s L}{\rho s} \tag{2.72}$$

being now s the distance to the ground line.

Figure 2.19 shows the frequency dependent conductance and capacitance, where the transition between slow wave and dielectric modes is clearly observed, for two cases: one corresponding to a nearby ground line, with $G_s \approx 5000\,\mathrm{S/m}$ and $C_p \approx 20\,\mathrm{pF/m}$, and the second case without a nearby ground line, where $G_s \approx 3\,\mathrm{S/m}$ and $C_p = 0$. As was to be expected, the addition of

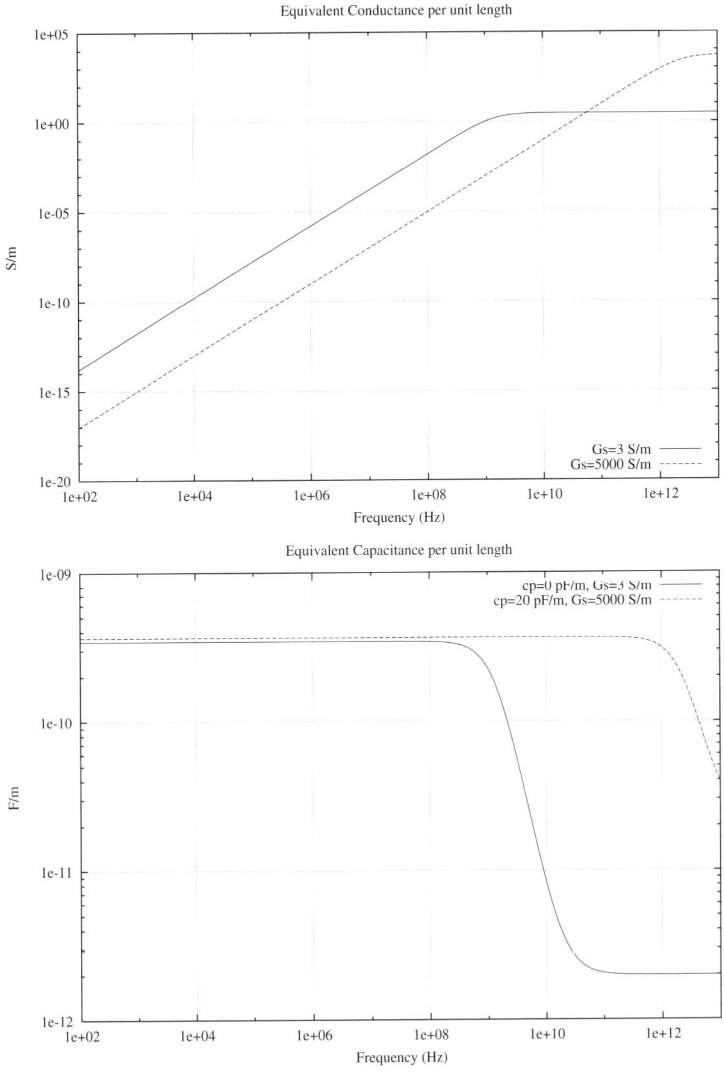

Figure 2.19. Frequency dependent conductance and capacitance per unit length, when no ground line is present ($G_s = 3\,\mathrm{S/m}$), and when there is a nearby ground line ($G_s = 5\,\mathrm{kS/m}$).

a nearby ground line tends to increase the range of frequencies for which the line has constant parameters (slow wave mode). In other words, with nearby ground lines, the influence of the semiconductor substrate becomes less important for transmission characteristics.

In summary, the existence of several modes translates to the usual $RLGC$ transmission line with frequency-dependent parameters. Recent advances in the simulation algorithms [12], [13], [14], have allowed the incorporation of such models in standard circuit simulation tools.

3.2 Example simulations of frequency-dependent parameters

Once obtained, the frequency variation of line parameters can be introduced in the description of a transmission line in modern circuit simulators, like Avant's HSPICE [15]. The parameter variation, however, becomes apparent only in the Giga-Hertz range for practical substrate and line technologies. Therefore, some examples are shown here to illustrate the difference between a constant-parameter model (RLC, as G is null at zero frequency), and one with varying parameters ($RLGC(f)$).

We will take two examples: one for a signal line close to ground return lines as shown in figure 2.17, and a second case with the same structure without the nearby ground lines, so the return is through the backplane.

The frequency dependent parameters C and G are those of figure 2.19, while L and R are calculated from Fasthenry [16] simulations (figure 2.20). In both cases, the small variation in inductance should be noted in the frequency range of DC up to 10 GHz. The DC value of inductance for the nearby ground line case is smaller than the other case, as should be expected. Taking the DC values of L and C, it is possible to calculate the ideal line characteristic impedance, $Z_o = \sqrt{L/C}$. As stated above in section 3, this magnitude must be compared to total line resistance and to driver resistance to determine the possibility of ringing in the line. For the two cases considered in this example, the values of Z_o are 48.77 Ω for the near ground line case, and 57.57 Ω for the case without a nearby ground line.

The simulations consist of comparing three models: a lossy RLC model with constant parameters, a frequency dependent $RLGC$ transmission line, and a simple lumped RC model. The driver for all models is a resistance of 50 Ω. This means that for the nearby ground case, $R_d > Z_o$, and therefore, no ringing can be observed. The opposite is true for the case without a nearby ground line, although the small difference between R_d and Z_o allows for only small oscillations.

The results are shown in figures 2.21 to 2.23. The general conclusion is that the presence of a nearby ground line makes the response of the constant- and frequency dependent- lossy transmission line models almost the same, for practical signals (the rise time considered was 15 ps). The importance of dispersion in the frequency dependent parameter model, as observed in the cases with no nearby ground line, is worthy of note. In those cases, dispersion leads to an apparent reduction of the propagation time: the voltage starts to increase, albeit slowly, very soon after the stimulus, similar to what happens with the RC model, also shown. It is also clear that the difference between both RLC

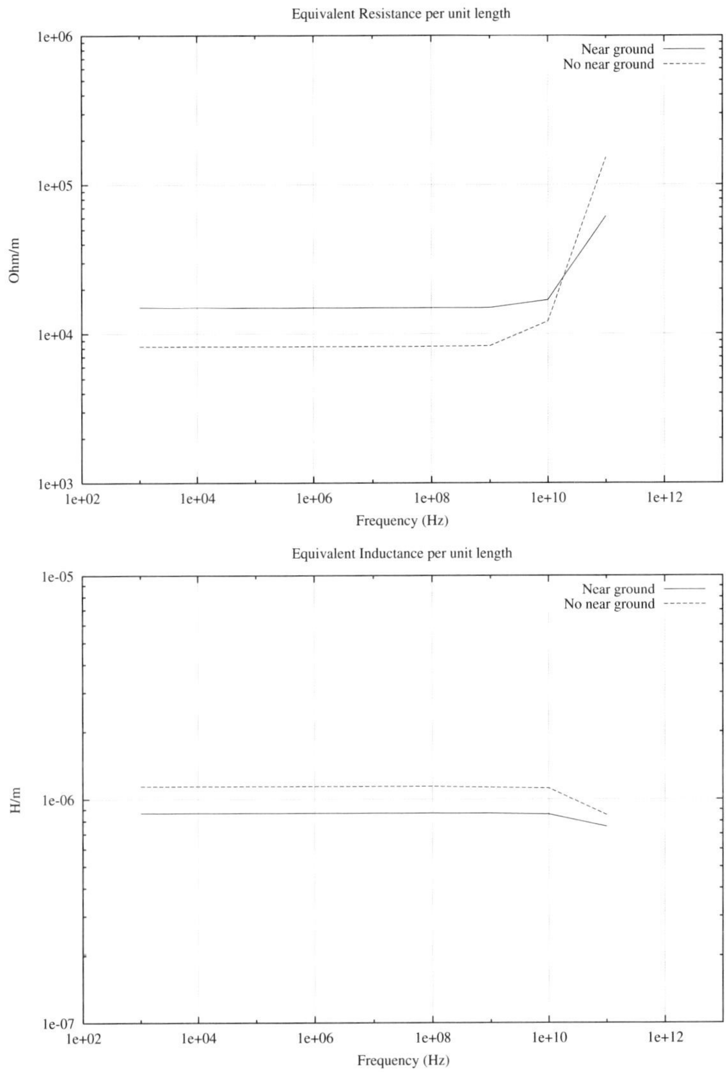

Figure 2.20. Frequency dependent resistance and inductance per unit length, when no ground line is present, and when there is a nearby ground line.

and $RLGC(f)$ models is apparent and significant only for long lines. In addition, for the cases where there is no ringing, the lumped RC model is a good approximation, particularly with respect to propagation delay.

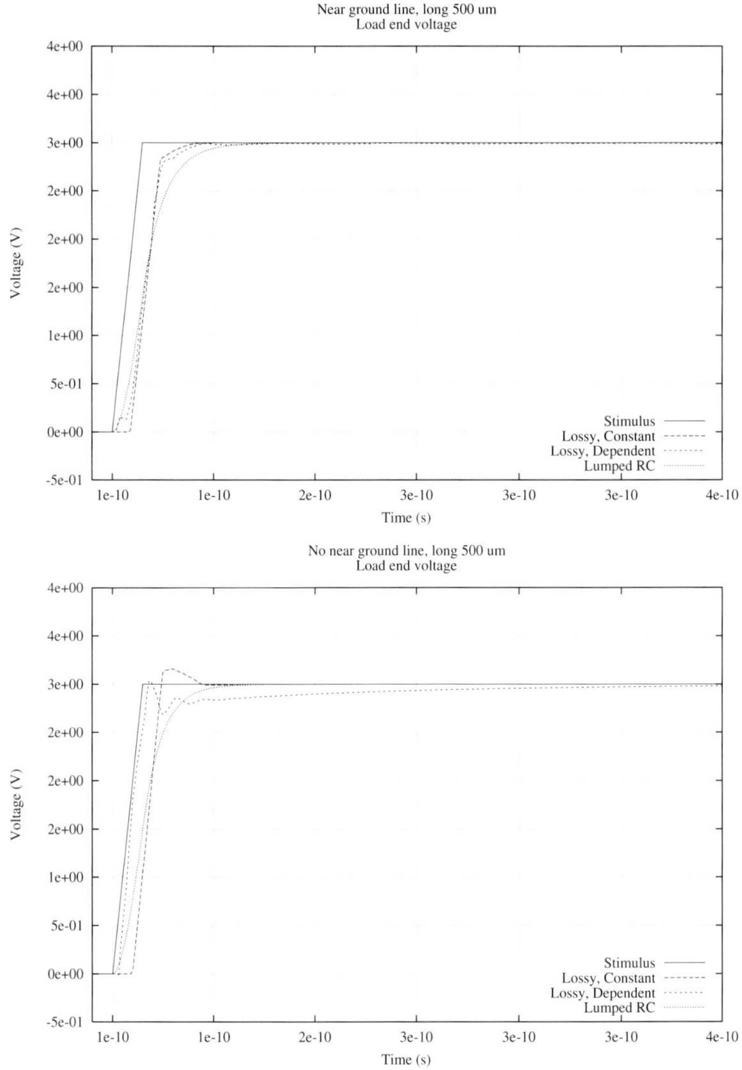

Figure 2.21. Transient response of Lossy transmission lines with constant and non constant parameters, for a line $500\mu m$ long.

4. Optimum line model selection

This chapter has presented different possibilities for electrically modeling an interconnection, together with the physics that justifies them. It has been shown that, in certain situations, there is almost no difference between the models. In other cases, the differences are not significant for the particular waveform property of interest (e.g. transition delay, or overshoot). These

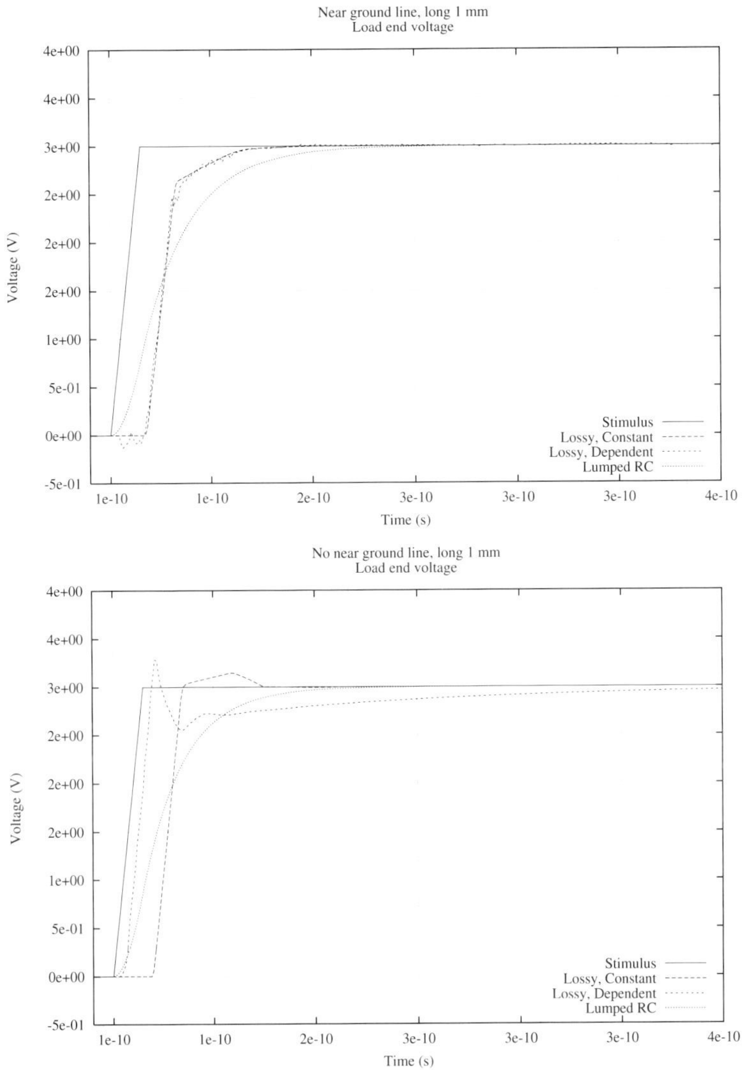

Figure 2.22. Transient response of Lossy transmission lines with constant and non constant parameters, for a line 1 mm long.

situations depend on line characteristics (parameters, line length), as well as driver and load characteristics, and on the frequency of the signals propagating through the interconnection. While the most complete model will always be more accurate, it is also true that the simulation of the complete model will take longer. Given the huge number of interconnections to be simulated in an integrated circuit, it is good to have an estimate for using the simplest model, which gives the response within a desired accuracy range.

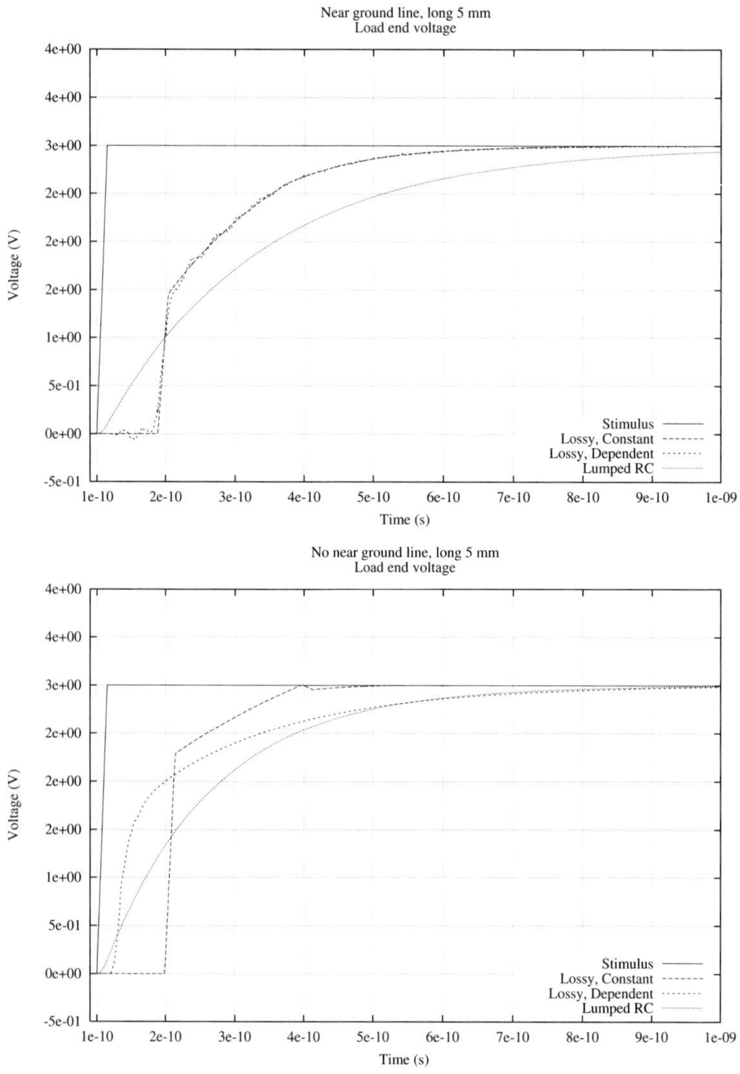

Figure 2.23. Transient response of Lossy transmission lines with constant and non constant parameters, for a line 5 mm long.

Following the classification of line models into inductive and non-inductive, as considered in section 1.3, we will use the above analysis of ringing as an orientation for deciding which electrical model to select. Depending on the input signal, the driver characteristics and line length, the response to a step voltage can be approximated as a non-inductive model (C or RC model) when there is no ringing, or as a more complex model including inductance (*RLC* or *RLGC* model). It is assumed here that parameters per unit length of the

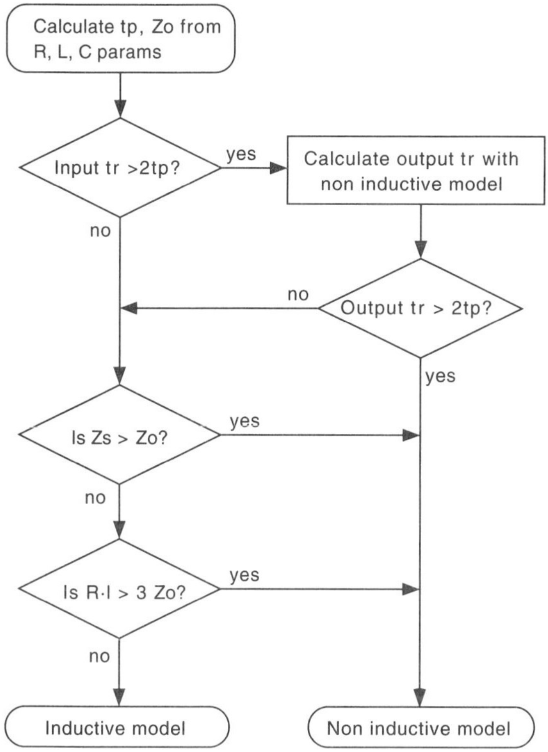

Figure 2.24. Process for the selection of an inductive or non-inductive line model.

line, R, L and C, are obtained from a layout analysis. The selection process is described in figure 2.24.

4.1 Effect of input rise time

The comparison between the signal's rise time and propagation time was shown to be important in section 2.2. However, the rise time to be compared is the one for the signal after the driver. In the case of a resistive driver model, it can be certain that the signal rise time at driver output will always be greater than the input rise time, but for active non-linear drivers, as in reality, this cannot always be assured.

Therefore, in general, the input rise time is not a definitive parameter for selecting the line model, although for cases when it is much greater than the propagation time, non-inductive models can be used to simulate and check the value of the rise time at driver output. If this approximate rise time is still greater than $2t_p$, then the inductive model can be discarded.

It must be noted that t_p depends proportionally on line length, l, as well as on the line's electrical parameters. From the velocity of propagation (expression 2.26):

$$t_p = l\sqrt{LC} \tag{2.73}$$

4.2 Effect of driver impedance

As shown in section 2, the driver impedance must be compared to the characteristic impedance of the ideal line for assessing the possibility of ringing, and therefore, the need for an inductive model. The usual capacitive interconnection termination of integrated circuits is assumed, so that $Z_o < Z_l$, and the condition for ringing is $Z_d < Z_o$. In general then, low impedance drivers will have greater chances of needing an inductive model for interconnection. The range of impedances in integrated circuits lies in the 50 to 300 Ω approximately [17].

4.3 Effect of line length

Line length influences model selection in two opposite directions. One is the already mentioned effect of propagation time: as length increases, reflections are more important (if the other parameters allow them to be). The other direction is related to line resistance: as line length increases, so does the total resistance, which therefore helps attenuate the possible oscillations. In summary, for inductive models to be necessary, two inequalities must be verified:

$$Rl < 2.5Z_o \tag{2.74}$$

$$t_r < 2l\sqrt{LC} \tag{2.75}$$

or:

$$\frac{t_r}{2}v_p < l < \frac{2.5Z_o}{R} \tag{2.76}$$

From this last expression, it can be concluded that the inductive model will be useful only for intermediate length lines, and that for some lines which verify $\frac{t_r}{2}v_p > \frac{2.5Z_o}{R}$, the non-inductive model is valid for any value of line length.

References

[1] Ramo S., Whinnery J. R., and Van Duzer T. *Fields and Waves in Communication Electronics*. Wiley and Sons, 2nd edition, 1984.

[2] Collin R. E. *Field theory of guided waves*. IEEE Press, 2nd edition, 1991.

[3] Montgomery C. G., Dicke R. H., and Purcell E. M., editors. *Principles of microwave circuits*. Dover, 1965.

[4] Mesa F. L., Cano G., Medina F., Marques R., and Horno M. "On the Quasi-TEM and Full-Wave Approaches Applied to Coplanar Multistrip on Lossy Dielectric Layered Media". *IEEE Trans. on Microwave Theory and Techniques*, 40(3), March 1992.

[5] Matthaei G. L., Shu J. C.-H., and Long S. I. "Simplified Calculation of Wave-Coupling Between Lines in High-Speed Integrated Circuits". *IEEE Trans. on Circuits and Systems*, 37(10), October 1990.

[6] Brews J. R. "Transmission Line Models for Lossy Waveguide Interconnections in VLSI". *IEEE Trans. on Electron Devices*, 33(9), September 1986.

[7] Horno M., Mesa F.L., Medina F., and Marques R. "Quasi-TEM analysis of multilayered, multiconductor coplanar structures with dielectric and magnetic anisotropy including substrate losses". *IEEE Trans. on Microwave Theory and Techniques*, 38(8), August 1990.

[8] DeFalco J. A. "Reflection and Crosstalk in Logic Circuit Interconnections". *IEEE Spectrum*, 7(7), July 1970.

[9] Ethirajan K. and Nemec J. "Termination techniques for high-speed buses". *EDN Magazine*, February 1998.

[10] Ho C. W., Chance D. A., Bajorek C. H., and Acosta R. E. "The Thin-Film Module as a High-Performance Semiconductor Package". *IBM J. Res. and Develop.*, 26(3), May 1982.

[11] Watts R.K., editor. *Submicron Integrated Circuits*, chapter 6, pages 269–331. John Wiley & Sons, 1989.

[12] Achar R. and Nakhla M.S. "Simulation of high-speed interconnects". *Proceedings of the IEEE*, 89(5), May 2001.

[13] Kuznetsov D..B. and Schutt-Aine J.E. "Optimal transient simulation of transmission lines". *IEEE Tran. on Circuits and Systems–I*, 43(2), February 1996.

[14] Odabasogliu A., Celik M., and Pileggi L.T. "PRIMA: Passive reduced-order interconnect macromodeling algorithm". *IEEE Trans. on Computer Aided Design of Integrated Circuits*, 17(8), August 1998.

[15] Avant! Corp. *Star-Hspice 2001.2*, June 2001.

[16] Kamon M., Tsuk M.J., and White J. "Fasthenry: a multipole-accelerated 3-D inductance extraction program". *IEEE Trans. on Microwave Theory and Techniques*, 42(9), September 1994.

[17] Moll F., Roca M., and Rubio A. "Inductance in VLSI interconnection modelling". *IEE Proc.-Circuits, Devices and Systems*, 145(3), June 1998.

Chapter 3

CROSSTALK EFFECTS IN DIGITAL CIRCUITS

This chapter deals with the description of the main effects that crosstalk coupling can introduce in integrated circuits. The coupling phenomenon is present in both digital and analog blocks. However, digital blocks are more densely interconnected and, for this reason, coupling tends to be more significant in this type of circuit. While the electrical models used in this chapter to model crosstalk effects can be also used to analyze analog circuits, this chapter analyzes it for digital circuits. The effects analyzed here are the first step in devising verification and test techniques for the crosstalk problem. This topic is addressed in chapter 6.

The circuit models used in the chapter are based on capacitive coupling. As shown in chapter 2, the inductance of the lines can be significant in some cases, but in integrated circuits, capacitance is the main parameter for describing interconnection response. The models used are very simple and this gives the advantage of providing an intuitive understanding of the problem that is useful for deriving design rules for preventing crosstalk.

One of the results of interconnect coupling in VLSI and ULSI circuits is the appearance of spurious (undesirable) signals in the coupled lines. This chapter starts with an analytical characterization of these spurious signals, showing their dependence on circuit parameters and their propagation capability throughout the circuit, together with measurement techniques.

A second crosstalk effect, very important in today's digital circuits, is the induced delay of transitions due to noise. This effect will be analyzed in depth in this chapter, including an extensive experimental measurement description.

Finally, the appearance of spurious signals in circuit lines can lead to changes in energy consumption due to these extra signals in the circuit. The implications of crosstalk noise in energy dissipation will be discussed through a simple coupling modeling of interconnect lines and drivers.

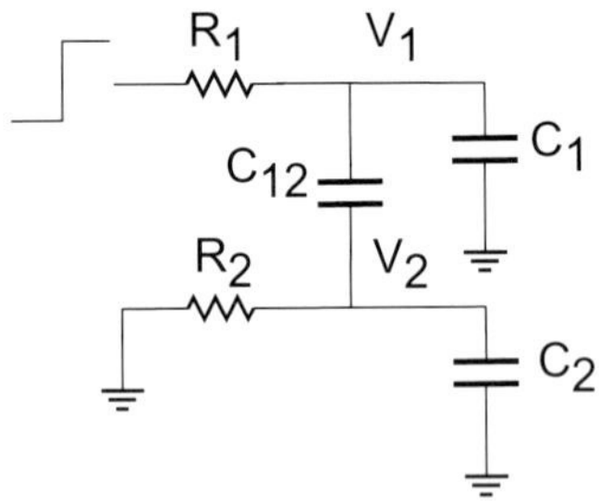

Figure 3.1. Circuit model for crosstalk between two lines.

1. Spurious signals

The increase in interconnection density may cause interference between neighboring lines. Electrically, this interference can be modeled through mutual capacitance, and in certain situations, also through mutual inductance. The phenomenon is called line coupling, or crosstalk.

When there is a voltage transition in one or more of several coupled lines, the quiet lines suffer a voltage perturbation which may be regarded as a spurious signal. In some cases, this spurious signal will be so small that it will produce no measurable effect. In other cases, it will be significant enough to affect the circuit's response. We can define the lines causing the perturbation affecting (where a normal voltage transition occurs) as the *affecting* or *culprit* lines, while the lines suffering the perturbation, where the spurious signal are generated, can be defined as the *affected* or *victim* lines.

In this section, we will analyze how the voltage waveform characteristics of a spurious signal depend on line and driver parameters, how this spurious signal propagates throughout the logic of the circuit and finally, how it can be measured.

1.1 Spurious signal characterization

In order to make a simple initial characterization of a spurious signal (SS) due to coupling noise in integrated circuits, let us consider the simple circuit model depicted in figure 3.1. In this scenario, one affecting line is modeled as an RC lumped line model, with R_1 being the driver resistance plus the line resistance, and C_1 being the line capacitance including load capacitance. In the same way, an affected line is modeled with a resistance R_2, and a capacitance C_2. Finally, C_{12} represents the coupling capacitance between the two lines. An ideal step input is assumed as the input of line 1, while line 2 is connected to zero through its driver resistance, R_2.

A detailed analysis of this model will provide us with some knowledge about how the different line and driver parameters affect the spurious signal produced, that is, how they influence the width and amplitude of the spurious signal. Considering the circuit in figure 3.1, the equations describing the waveform in the affecting line are [1]:

$$V_1(t) = V_a(t) + V_b(t) + V_c(t) \tag{3.1}$$

where

$$V_a(t) = \frac{\frac{1}{R_1 R_2 C_T} + \frac{(C_{12}+C_2)s_1}{R_1 C_T}}{s_1(s_1 - s_2)} e^{s_1 t} \tag{3.2}$$

$$V_b(t) = \frac{\frac{1}{R_1 R_2 C_T} + \frac{(C_{12}+C_2)s_2}{R_1 C_T}}{s_2(s_2 - s_1)} e^{s_2 t} \tag{3.3}$$

and

$$V_c(t) = \frac{\frac{1}{R_1 R_2 C_T}}{s_1 s_2} u(t) \tag{3.4}$$

being $u(t)$ the unitary step function:

$$u(t) = \begin{cases} 0 & \text{for } t < 0 \\ 1 & \text{for } t \geq 0 \end{cases} \tag{3.5}$$

For the affected line, the result can be expressed as:

$$V_2(t) = \frac{C_{12}}{R_1 C_T (s_1 - s_2)} (e^{s_1 t} - e^{s_2 t}) \tag{3.6}$$

In the above expressions, C_T is:

$$C_T = C_{12}C_2 + C_{12}C_1 + C_1 C_2 \tag{3.7}$$

and s_1 and s_2 are the roots of the following second-order algebraic equation:

$$s^2 + \frac{R_1(C_{12} + C_1) + R_2(C_{12} + C_2)}{R_1 R_2 C_T} s + \frac{1}{R_1 R_2 C_T} = 0 \tag{3.8}$$

The amplitude of the spurious signal produced in line 2 is obtained calculating the maximum of V_2 by evaluating equation 3.6 at $t = t_m$, being t_m:

$$t_m = \frac{log(s_1/s_2)}{(s_2 - s_1)} \tag{3.9}$$

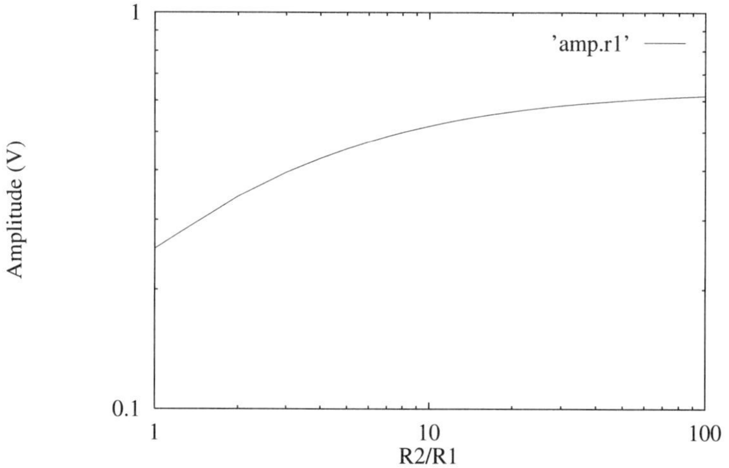

Figure 3.2. Spurious signal amplitude as a function of total node resistance ratio.

The width of the crosstalk waveform (evaluated at half amplitude) must be calculated numerically, because no analytical expression for this magnitude can be found.

By applying the above expressions, it is possible to analyze the influence of the different circuit parameters such as driver sizes, line configuration, line loads, etc. Figures 3.2 and 3.3 show how the amplitude and width depends on the ratio between victim and affecting total node resistance (including both driver and line resistances). The calculations presented are made considering two lines with 170 fF of ground capacitance each, and 300 fF of mutual capacitance between them. The resistance of the victim line (R_2) considered is 1 kΩ. As can be seen, the amplitude increases asymptotically with the ratio between driver resistances, tending to a value $\Delta V_{2,\max}$ determined by the following capacitance relation:

$$\Delta V_{2,\max} = \frac{C_{12}}{C_{12} + C_2} \tag{3.10}$$

It is also shown that for equal driver resistances ($R_2/R_1 = 1$) the value of the amplitude is quite small, about 40% of the value predicted by expression 3.10. (Note that in all the cases depicted, the signal amplitude in the affecting line is assumed to be 1 V, in other words, the amplitude of all figures is normalized).

In figures 3.4 and 3.5, the amplitude and width of crosstalk spurious signals are computed in terms of victim resistance driver. The values of capacitances are in this case 170 fF for both ground and crosstalk capacitances. Node re-

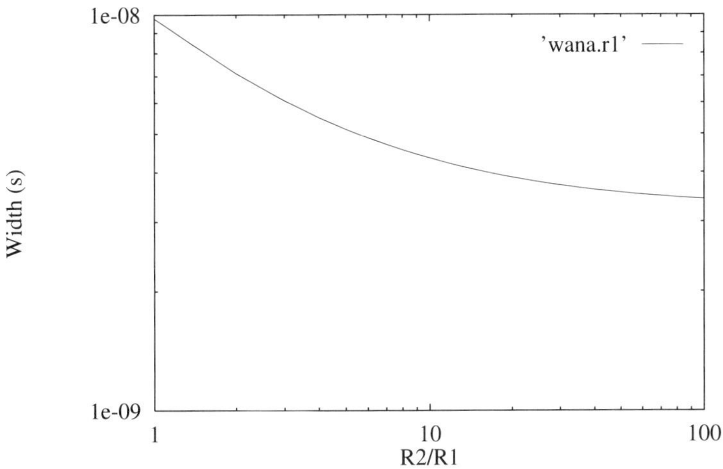

Figure 3.3. Spurious signal width as a function of total node resistance ratio.

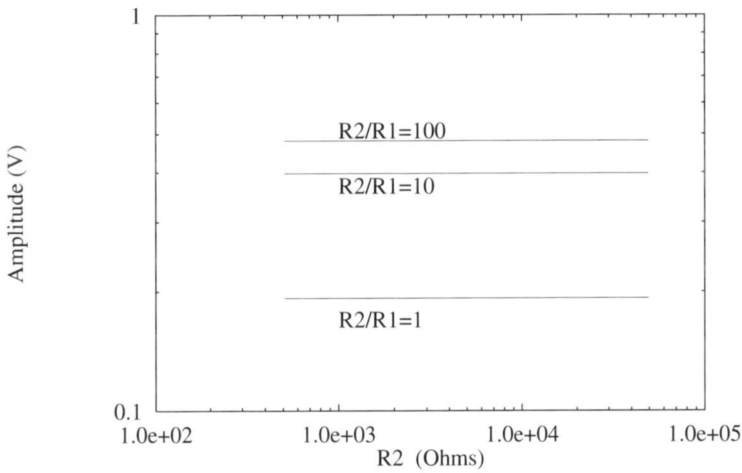

Figure 3.4. Spurious signal amplitude as a function of the victim node resistance for fixed culprit–victim resistance ratio (1, 10 and 100).

sistance ratios of 1, 10 and 100 are considered. It is shown that the amplitude is independent of the value of resistance and only depends on the resistance ratio. In addition, the signal width increases with the affected node resistance, as may be intuitively expected because the victim line RC constant increases.

We will study now the dependence of amplitude and width in function of the capacitance ratio ($\frac{C_{12}}{C_{12}+C_2}$). Figures 3.6 and 3.7 show the plots of these magnitudes for different values of coupling capacitance to affecting line ca-

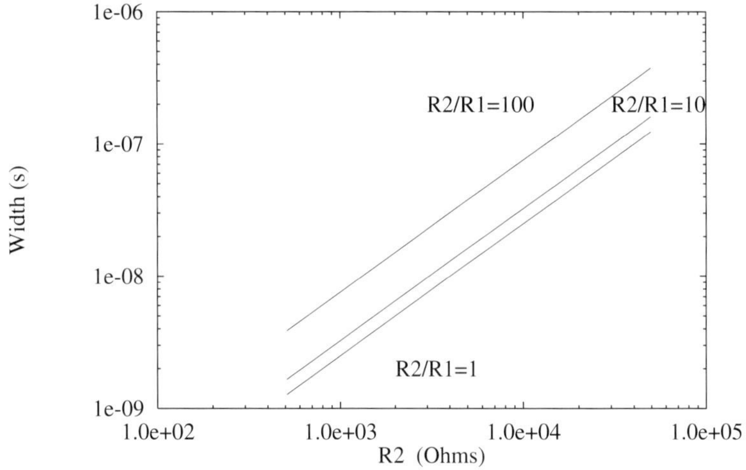

Figure 3.5. Spurious signal width as a function of the victim node resistance for fixed culprit–victim resistance ratio (1, 10 and 100).

pacitance ratio ($\frac{C_{12}}{C_1}$), and taking equal node resistances. It can be observed that the crosstalk amplitude increases with the x axis as can be expected from expression 3.10, and also with the ratio between coupling and affecting line's ground capacitances. On the other hand, the width of the spurious signal decreases as the ratio $\frac{C_{12}}{C_{12}+C_2}$ increases. It is important to note in this case that the amplitude never reaches the usual logic threshold voltage of digital CMOS circuits ($V_{DD}/2$), independently of the value of capacitances. Therefore, a design rule for averting potential problems in logic circuits is to try to balance the resistance of coupled lines.

Figures 3.8 and 3.9 show the same results as figures 3.6 and 3.7 for the case where the driver resistance of the victim line is 10 times bigger than the driver resistance of the affected line. Similar trends are obtained as in the case of equal driver resistances, but now greater values of amplitude and width are obtained. In this case, the amplitude can reach values bigger than usual logic threshold voltage of digital CMOS circuits.

Finally, this simple coupling line model allows the possibility of analyzing the effect of the line length in the crosstalk phenomenon. In this case, the node's resistance and capacitance are explicitly put in terms of the parameters of the line and the driver as:

$$R_1 = R_{1l}l + R_{1d} \tag{3.11}$$
$$R_2 = R_{2l}l + R_{2d} \tag{3.12}$$
$$C_1 = C_{1l}l + C_{1r} \tag{3.13}$$

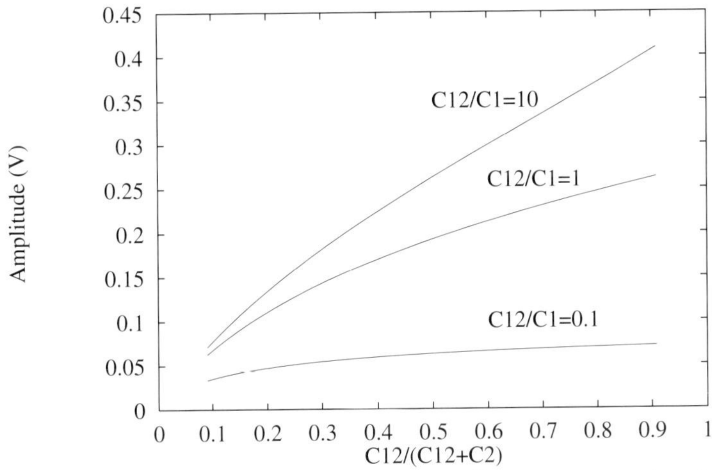

Figure 3.6. Spurious signal amplitude in function of $\frac{C_{12}}{C_{12}+C_2}$ for equal node resistances.

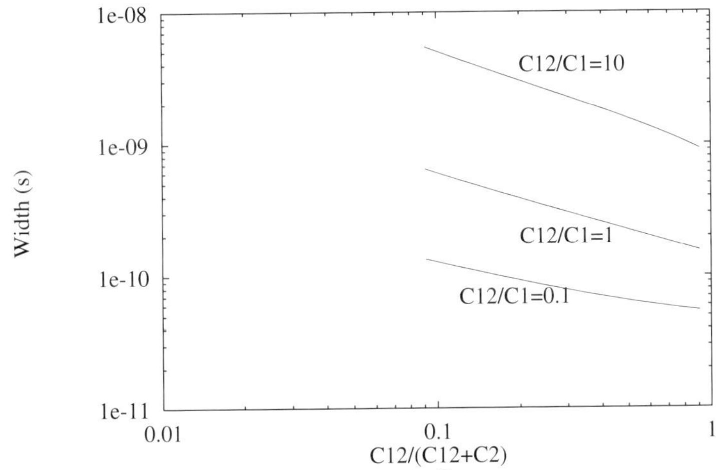

Figure 3.7. Spurious signal width in function of $\frac{C_{12}}{C_{12}+C_2}$ for equal node resistances.

$$C_2 \quad = \quad C_{2l}l + C_{2r} \tag{3.14}$$

where R_{1l}, R_{2l}, C_{1l} and C_{2l} are line parameters, l represents the line length, R_{1d} and R_{2d} are driver resistances and C_{1r} and C_{2r} are load capacitances of the lines.

It can be shown [1] that when R_{1d}, R_{2d}, C_{1r} and C_{2r} are null, so that the resistance and capacitances are strictly proportional to length, the crosstalk amplitude is independent of length and the width is quadratic with length.

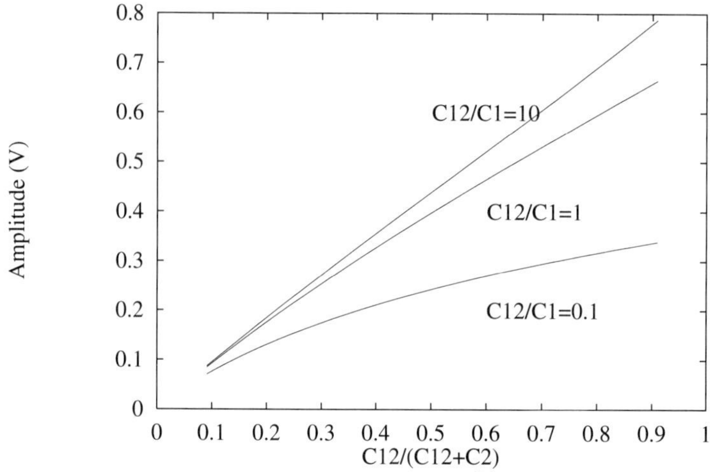

Figure 3.8. Spurious signal amplitude in function of $\frac{C_{12}}{C_{12}+C_2}$ for a resistance ratio of 10.

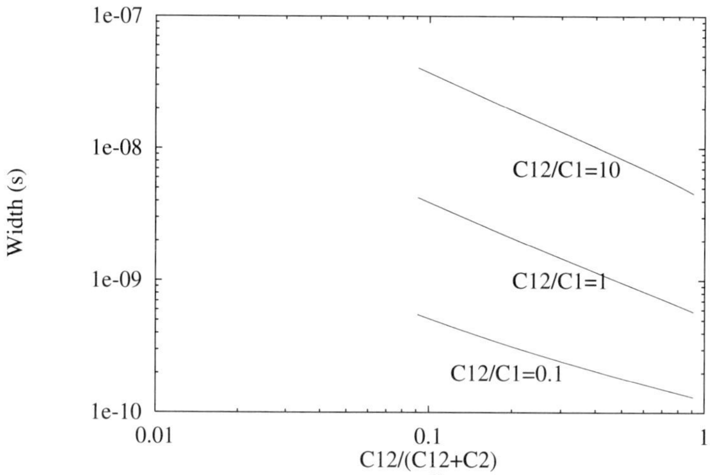

Figure 3.9. Spurious signal width in function of $\frac{C_{12}}{C_{12}+C_2}$ for a resistance ratio of 10.

The situation changes when a non-null driver and load parameters are considered. This case is shown in figures 3.10 and 3.11, for two different values of the relation of driver resistances $\rho = \frac{R_{2d}}{R_{1d}}$. It is shown that for increasing values of ρ, greater values of amplitude and width are obtained, in accordance with the results of figures 3.2 to 3.5. An interesting result obtained in this case is the appearance of a maximum in the curves for a large driver resistance ratio, indicating that for a given case it exists a value of length implying maximum amplitude of crosstalk. When line length increases, coupling capac-

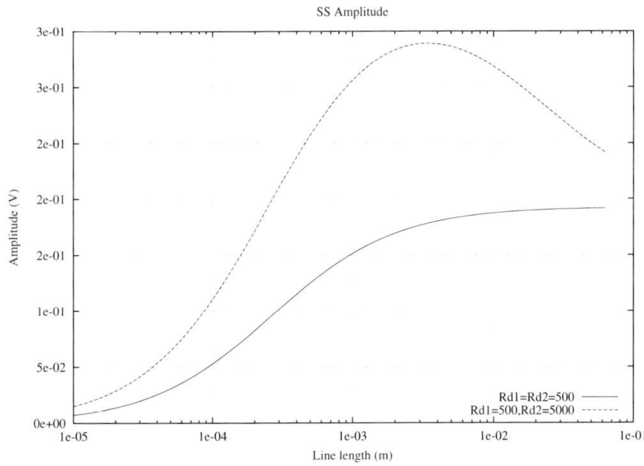

Figure 3.10. Spurious signal amplitude as a function of line length for several driver resistance ratios.

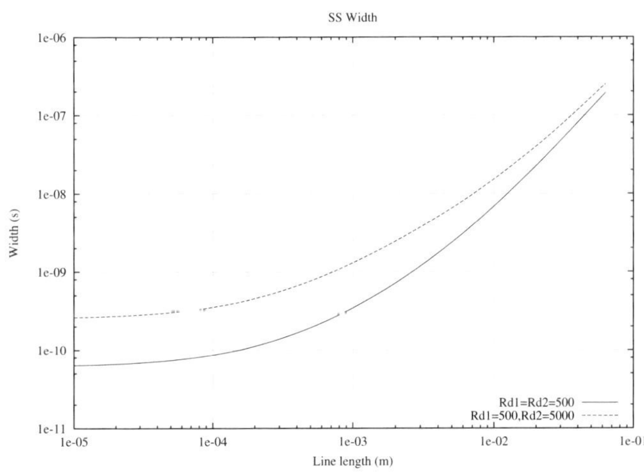

Figure 3.11. Spurious signal width as a function of line length for several driver resistance ratios.

itance increases, but the rise time of the affecting signal is increased due to the increased RC constant of the affecting line; that is, two opposite factors play an important role in the crosstalk noise phenomenon and this explains the maximum observed in these figures.

1.2 Propagation of spurious signals through VLSI digital circuits

The effect of spurious signals on the logical behavior in digital circuits will be analyzed here. Depending on the amplitude and duration of the spurious signal, the subsequent digital gate may interpret it as an actual valid signal, and then a logic error is produced and propagated. This error can be permanent if it affects the circuit's sequential block. In combinational blocks, only transient errors will appear.

The main waveform parameters influencing the effect of spurious signals through logic are its amplitude and its width. In order to cause a logic error in a digital gate, the signal amplitude must be higher than the logic threshold voltage (switching voltage) of the affected gate (the gate with the spurious signal at its input). If the amplitude does not reach this value, a logic error cannot be generated, and for this case we do not consider spurious signal propagation through the gate.

When the amplitude is higher than the switching voltage of the affected gate, then the possibility of propagation of the signal will depend on its width and also on the propagation delay of the affected gate. The study of the propagation capability of a signal is addressed in the next section.

1.2.1 Penetration depth of a spurious signal

The fact that the propagation depends on both amplitude and width makes it natural to characterize any spurious signal as a square voltage pulse.

In order to evaluate the propagation capability of a given signal, we present the following definition [2]:

DEFINITION 3.1 *The penetration depth of a spurious signal is defined as the number of gates that the signal can pass through, causing a logic error at the last one.*

Now, the characterization, or the process of assigning a square voltage pulse to a given spurious signal, can be made using two different criteria:

- To choose an equivalent pulse that has a propagation effect equal to the spurious signal considered. The main problem of this criteria is that the effect of the spurious signal is unknown, therefore it is impractical.

- To find an equivalent pulse with a penetration depth that is an upper or lower limit of the spurious signal's penetration depth.

The following definition will be used to find upper and lower limits of a spurious signal's penetration depth:

DEFINITION 3.2 *A given signal A is said to cover a signal B if, after a certain time shift of one of them, it is verified that:*

$$|\Delta V_B(t)| \leq |\Delta V_A(t)| \tag{3.15}$$

for all the time interval during which $|\Delta V_B(t)| \geq V_{th}$, where $|\Delta V_B(t)|$ and $|\Delta V_A(t)|$ are the moduli of voltage variation of both signals compared to their initial steady state value, and V_{th} is the threshold voltage of the MOS transistors.

Considering that the incremental transient output voltage function of an electronic logic device follows a monotonic behavior with respect to the incremental input voltage [3], the following statement can be made:

LEMMA 1 *Given a signal A covering a second signal B, and their output waveforms from a certain logic gate at which both signals arrive being $O(A)$ and $O(B)$, respectively, it is verified that signal $O(A)$ covers signal $O(B)$ as well.*

From this, it is possible to see that a signal A covering a second signal B will have a propagation depth (number of logic gates that the signal can actively traverse) greater than or equal to that of signal B.

Now, given a certain spurious signal SS, we can define two sets of square voltage pulses: one set (US) consisting of the pulses covering SS, and a second set (LS) composed of the pulses covered by SS. From the above statement, it is easy to see that the penetration depth of any signal from the set US is an upper limit for the spurious signal's penetration depth, while the penetration depth of any signal LS is a lower limit. Figure 3.12 shows the waveforms of an arbitrary spurious signal, US and LS signals.

1.2.2 Example: penetration depth of a spurious signal through a chain of CMOS inverters

As an example of this formulation, we consider now an unlimited chain of identical CMOS inverters as is depicted in figure 3.13, each of them loaded by identical equivalent capacitances. Then, given a square pulse of amplitude H and width W, we calculate the maximum number k of inverters such that the spurious signal produces an incremental signal with amplitude above the logic threshold (assumed $V_{DD}/2$) at the output of the last one. In relation to this problem, the following definitions can be given.

DEFINITION 3.3 *We define k-level critical pulse as an ideal square pulse with amplitude H and width W such that it produces a signal of amplitude equal to the logic threshold value at the output of the k-th stage.*

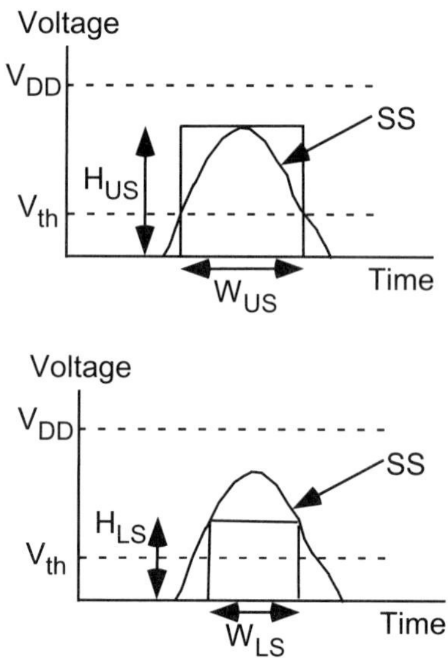

Figure 3.12. SS waveform and US and LS signals.

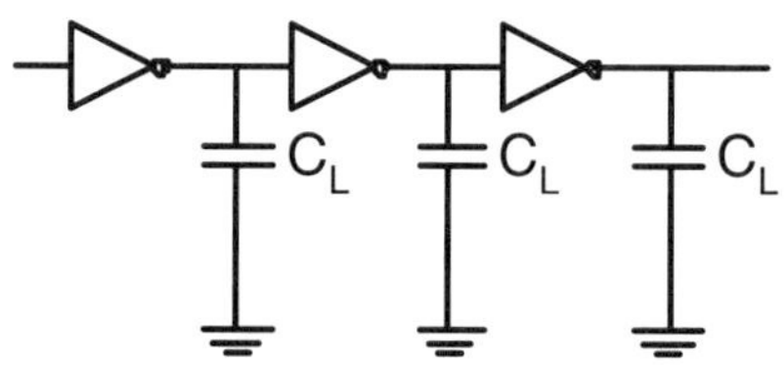

Figure 3.13. Chain of inverters with equal load capacitance.

DEFINITION 3.4 *We define k-level penetration critical curve as the relationship between amplitude H and width W of all the k-level critical pulses.*

Thus, the calculation of pulse penetration depth consists of calculating all k-level penetration critical curves, for k ranging from 1 to the number of gates of the path the spurious signal (SS) has to follow to the output. For each pulse amplitude value, H, there is a set of pulse widths $W^{(k)}$, corresponding to the k-level critical pulses. Each $W^{(k)}$ corresponds to a solution of the equation

$$W^{(k)} = F^{(k)}(H) \tag{3.16}$$

with $F^{(k)}$ depending only on technological parameters.

This calculation allows an analytical treatment for $k = 1$ alone, with the MOS devices of the inverter (PMOS and NMOS transistors) being modeled using the Sah model [4]. The solution for $W^{(1)}$ in function of H is given by the following expressions [5].

- Case $V_{DD}/2 < H < V_{DD}/2 + V_{th}$:

$$W^{(1)} = \frac{C_L}{K_{p,n}} \left(\frac{2}{V_S} arctan \left[\frac{V_H}{V_S} \right] + \frac{4(V_{DD}/2 - V_H)}{V_S^2} \right) \tag{3.17}$$

- Case $V_{DD}/2 + V_{th} < H < V_{DD} - V_{th}$:

$$W^{(1)} = \frac{C_L}{K_{p,n}} \left(\frac{2}{V_S} arctan \left[\frac{V_H}{V_S} \right] + \frac{4V_{th}}{V_S^2} + \frac{1}{V_S} log \left[\frac{V_S + V_L - V_{DD}/2}{V_S - V_L + V_{DD}/2} \right] \right) \tag{3.18}$$

- Case $H > V_{DD} - V_{th}$:

$$W^{(1)} = \frac{C_L}{K_{p,n}} \left(\frac{2(V_{DD} - V_L)}{V_L^2} + \frac{1}{V_L} log \left[\frac{2V_L - V_{DD}/2}{V_{DD}/2} \right] \right) \tag{3.19}$$

where $V_H = V_{DD} - V_{th} - H$, $V_L = H - V_{th}$, $V_S = \sqrt{V_L^2 + V_H^2}$, and V_{DD} the power supply voltage, V_{th} the threshold voltage (assumed the same in NMOS and PMOS), $K_{p,n}$ the transistor transconductance (the same for PMOS and NMOS in balanced inverters), and C_L the loading capacitance of each inverter in the chain.

Figure 3.14 shows the 1-level critical curve, where the width axis W is normalized by $C_L/K_{p,n}$. The points to the left of this curve cause no logic response at the output of the first inverter, while the points to the right are able to cause a signal of amplitude at least above the logic threshold at the output of the first stage. This curve is related to the well-known concept of the dynamic noise immunity of a logic device, and its analytical equation depends on technology and design parameters.

An extension to an arbitrary value of k can be made using numerical methods to solve the equations. Figure 3.15 shows the results of the calculation of the k-level penetration critical curves for different values of k, and with the same time normalization as indicated above. As can be seen in the graph, for increasing values of k, the difference between the curves becomes smaller and smaller.

1.3 Measurement of spurious signals

In order to evaluate crosstalk models correctly it is important to be able to measure the signals produced because of crosstalk. However, this phenomenon is very difficult to characterize experimentally because it is produced at an internal node and its effect may or may not be transmitted to an external output.

A first measurement approach consists of designing the circuit with metal pads connected to internal nodes, and then using contact probes to measure the crosstalk waveforms directly. However, this is not a good solution because

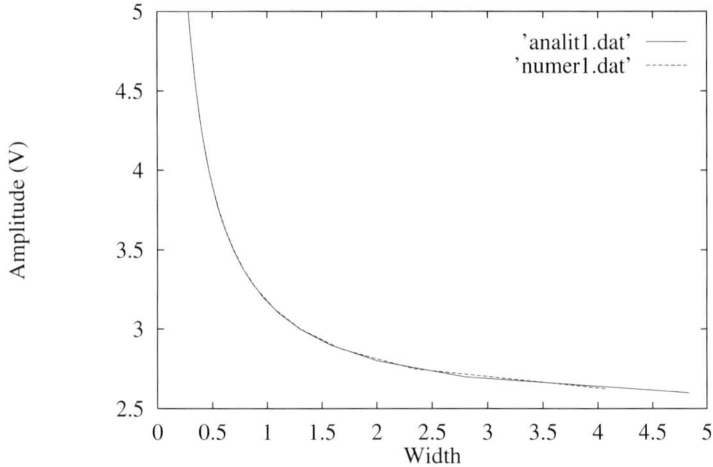

Figure 3.14. Critical curve for k=1. Width is expressed in normalized units of $C_L/K_{p,n}$.

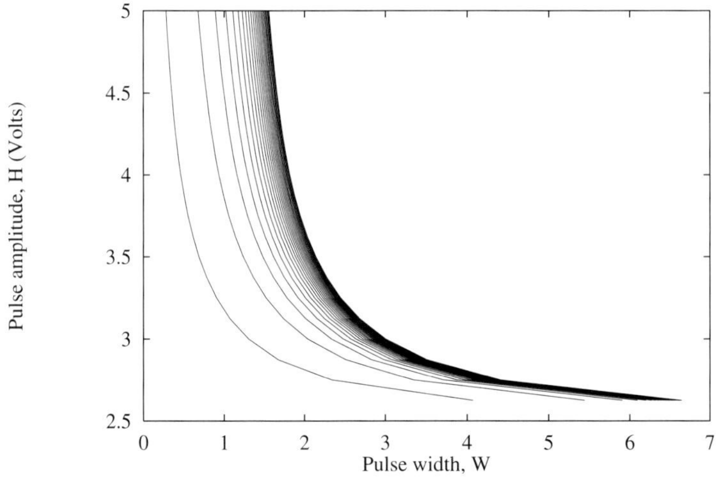

Figure 3.15. k-level penetration critical curves. Width is expressed in normalized units of $C_L/K_{p,n}$.

the inclusion of pads will dramatically modify the physical interconnection structure, leading to important changes in the values of the lines' parasitic capacitance. Important changes are therefore introduced in the crosstalk phenomenon.

Other solutions include noise measurements at the external pins of the circuit. In this case, packaging and bonding effects can alter the noise as well, so this kind of methodology is not suitable for determining crosstalk noise in integrated circuits experimentally.

Another approach consists of an integrated diagnosis system using an electron beam microscope based on secondary electron reflections from the metal lines on the circuit. However, this method can not be used for precise crosstalk characterization. The reason is that crosstalk noise is masked by other noise coming from oxide diffraction and also because only upper layer measurements can be made with acceptable accuracy. Lower metal lines are screened by the presence of upper layers.

For all of these reasons, it seems that indirect measurements are necessary for evaluating noise inside integrated circuits. These indirect measurements consist of analyzing the effect of noise in well-known circuit blocks, or in performing internal sampling strategies. J. Cartrysse, in [6] presents a method for measuring waveform distortion of an opamp due to substrate coupling. The same author, in [7] describes crosstalk measurement techniques through analyzing the effect of crosstalk in a latch structure.

The problem is similar when the objective is to characterize waveform propagation in the interconnection structure of silicon technology. With this objective, Soumyanath et al. published in [8] an interesting measurement approach to accurately evaluating on-chip interconnect behavior. In this work, a delay measurement system is developed based on a rail-to rail comparator applied to a set of several lines. The comparators are connected to the beginning of the lines and to the end of the lines. Then, by comparing the results obtained at the beginning with the results obtained at the end, time domain characteristics can be extracted. This technique is applied to a technology including five metal levels.

In [9], [10] and [11], another strategy is presented, based on a sampling sensor. A sampling transistor is delayed with respect to an affecting signal, so that each time a different point of the spurious signal is sampled (it is assumed that the spurious signal is exactly the same at each crosstalk event). An amplifier in follower configuration gives the sampled data at its output. When this methodology is applied, a reconstruction task is essential in order to obtain the voltage waveform.

Yet another approach, focused on measuring noise peak and delay changes, can be found in [12].

In this section, we will present a sensor for on-chip crosstalk measurements between coupling metal interconnects based on the indirect effect of crosstalk on a latch structure, in a similar way to [7]. The application of this sensor to several structures and integrated technologies can be found in [13]. The sensor provides precise crosstalk noise amplitude measurement not altered by test pads or probes.

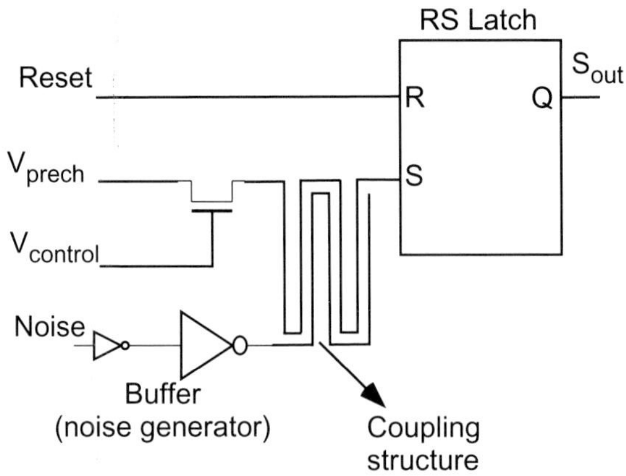

Figure 3.16. RS sensor architecture.

1.3.1 Sensor structure and operation

The sensor architecture is shown in figure 3.16, where three different parts can be distinguished:

- Noise generator: this is constructed using a tapered inverter chain in order to obtain a sharp edge to the source signal. The last inverter has a high fan-out capability in order to induce strong coupling.
- RS latch: it is constructed with NAND gates to measure inverted spurious signals, and with NOR gates for non-inverted ones. The latch switching voltage is denoted by V_c. The value of V_c can be determined by properly designing the NOR gate (NAND gate) transistor dimensions.
- Pre-charge circuit: this is constructed using an MOS pass transistor. This pass transistor is an n-channel MOS for NOR RS to have an accurate pre-charge of the set line in the range 0 to $V_{DD} - V_{TN}$. For a NAND RS, the pass transistor is a p-channel MOS. The selection of one or another depends on the direction (inverted or non-inverted) of the spurious signal to be detected.

The procedure for measuring the crosstalk noise (figure 3.17) is as follows: first, the RS latch is reset (time t_0), and after that, the NMOS pass-transistor is turned on (by activating $V_{control}$ at time t_1). Consequently, the *Set* line (victim line) is pre-charged to an externally fixed analog value V_{prech}, chosen to be lower than the latch switching voltage V_c. It is possible with this circuit to modulate the equivalent driver resistance in the victim line: if the pass transistor is cut off, the victim line shows a very high impedance and, depending on the value of $V_{control}$ and the pass transistor dimensions, the impedance of

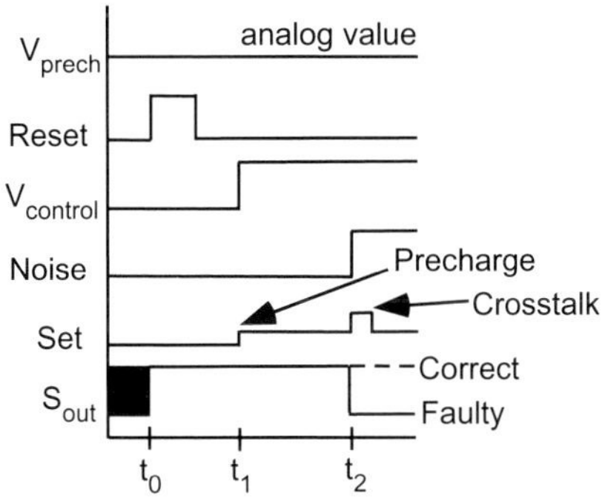

Figure 3.17. Coupling measurement in the victim line.

the line is lower. From the analysis in section 1.1, the ratio between node resistances has a strong influence on crosstalk noise characteristics. After that, the noise generator is activated by inducing a sharp transition in the source line (time t_2). Due to crosstalk, this transition produces a noise peak that pulls up the *Set* line. If the noise is strong enough to put the *Set* line to a voltage beyond the commutation point V_c, then the output sensor (S_{out}) switches to an erroneous state. Let V_{lim} be the limit pre-charge voltage such that the RS latch is not affected by noise. The voltage difference $V_c - V_{lim}$ will then be the crosstalk noise amplitude.

As an example of measurement application of the sensor, we mention here some results obtained with BiCMOS technology. More examples can be found in [13], where an exhaustive set of different sensors applied to different technologies is presented. The example here considered deals with the comparison between simple and multiple crosstalk, that is, what is the effect of crosstalk when the affecting line is a line near the victim, or the affecting lines are two lines, at either side of the victim line. The sensor architecture considered is based on RS NAND sensors, so the objective is the detection of inverted spurious signals. The noise generator is made of three buffered inverters. The coupling structure is composed of three lines at the first metal level, of minimum width and with minimum distance between them. The central line is the victim, while the adjacent lines are the culprits. A microphotograph of the integrated circuit can be seen in figure 3.18.

The measurements have been developed in the following way. First, the switching voltage of the RS latch (V_c) has been measured. Once this value has been obtained, the procedure for coupling measurement has been applied, as described above. Different line impedances have been considered (by changing the voltage at the gate of the PMOS pass transistor). Two different cases have been measured, first with only one affecting line switching, and second when the two affecting lines make simultaneous transitions. The results are depicted

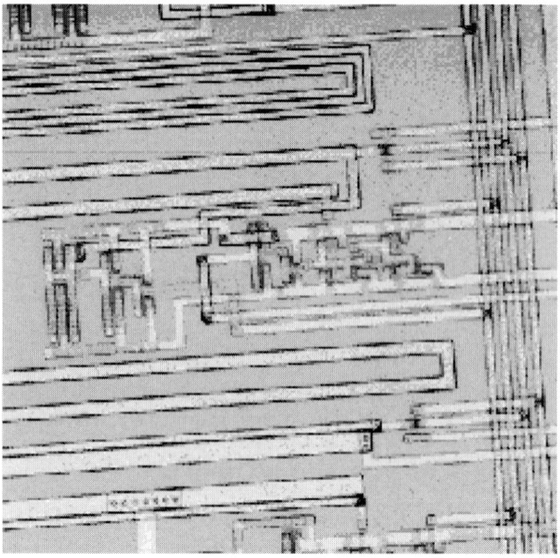

Figure 3.18. Microphotograph, a detail of RS NAND sensors and the buffered inverters.

in figure 3.19. The y-axis is the difference between V_{lim} and V_c, that is, the crosstalk noise amplitude measured, while the x-axis represents the impedance of the equivalent driver of the victim line. This value of impedance is the equivalent impedance of the pass transistor, which is in linear region for all the measurements performed. Thus, this impedance is approximately given by the expression:

$$R = \frac{1}{K_p \left(V_{SG} - |V_{TP}| \right)} \tag{3.20}$$

where K_p is the transconductance of the transistor, and can be expressed as:

$$K_p = \mu_p \frac{\epsilon_{ox}}{t_{ox}} \frac{W}{L} \tag{3.21}$$

The parameters of the last equation are given by the foundry, except W and L, which corresponds to the width and length of channel transistor, fixed by our design. It is observed in the figure that simple crosstalk represents about 50% of double crosstalk. A similar experiment has been performed by changing the role of the different lines. This means considering that a surrounding line is the victim and the central one plus the other surrounding the victims. From this measurement we have obtained the fact that the greatest effect is due to the central line and we can conclude that in crosstalk noise analysis only neighboring lines need to be considered [13].

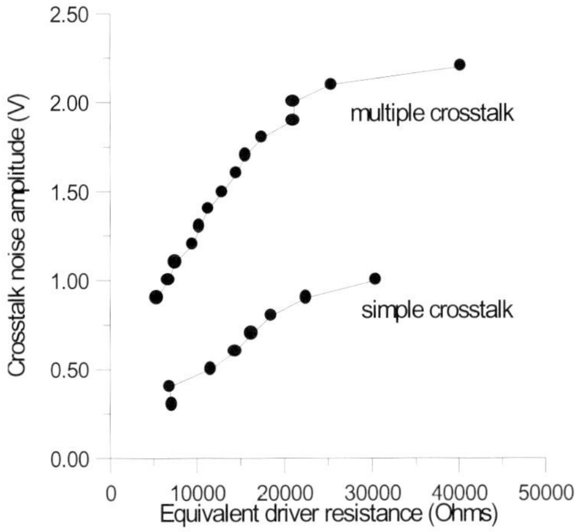

Figure 3.19. Crosstalk spurious signal amplitude measurements in a BiCMOS technology.

2. Crosstalk induced delay

The problem of spurious signals, dealt with in the previous section, is not the only one that can be produced by crosstalk. A second very important effect is the introduction of changes into signal propagation delay. This effect appears when the two coupled lines are switching simultaneously. This case can be understood as a spurious signal superimposed on the switching victim signal. Thus, a change in delay of the propagated signal is observed.

Figure 3.20 shows a simulation of this effect. The upper signal represents the waveform simulated at the affecting line, a low-to-high (LH) transition in the node. Almost simultaneously, the victim line is making a transition in the opposite direction (HL). The signal at the bottom of the diagram represents the waveform at the victim line when no coupling between lines is considered. The middle waveform corresponds to the waveform obtained through simulation in the victim line when capacitive coupling between lines is considered. A positive spurious signal added to the switching waveform of the victim line can be observed, which implies an increase in the effective delay of the victim line (δ). Actually, in the case of simultaneous transitions, both lines are at the same time victims and culprits, and the first line in the diagram will also show a variation in transition delay due to crosstalk. As happened with spurious signals, the least resistive node (line plus driver) will be the least affected and the most resistive node will be the most affected.

It must be noted that this delay variation can be positive or negative, depending on the direction (LH or HL) of the two simultaneous transitions. If both transitions have the same direction, the propagation delay for both transitions is decreased and if transitions are in opposite directions, delay is increased. An

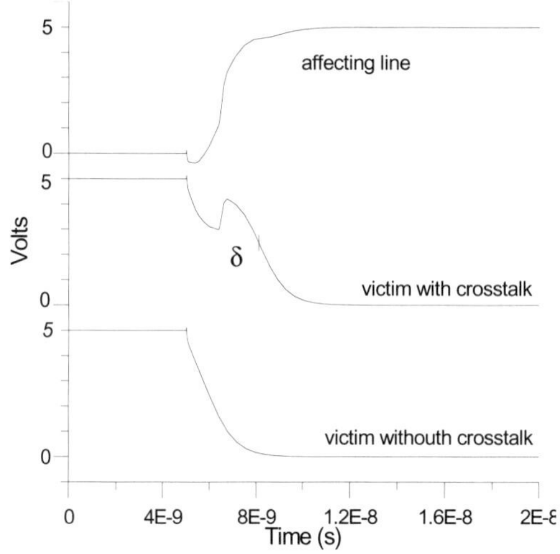

Figure 3.20. Simulated crosstalk induced delay.

important factor determining the magnitude of this effect is the synchronism between transitions. In addition, the width of the spurious signal responsible for the delay variation is more important than its amplitude. Thus, highly resistive lines generating broad spurious signals can be very susceptible to this effect. The consequence of all this is that spurious signals with amplitude smaller than the switching voltage of the logic can have important effects on the behavior of the circuit, in contrast to what happens with pure propagation of spurious signals through logic.

The influence on the behavior of the circuit will depend on the value of the additional delay. This value will depend on various factors, some of them the same as those influencing spurious signals, plus others, such as the synchronism between transitions. This effect can be very important in sequential synchronous circuits where erroneous logic values can be stored due to timing crosstalk problems.

2.1 Experimental measurement approaches

For the same reasons mentioned in the experimental measurement of spurious signals, it is not easy to perform direct measurements in order to obtain waveforms similar to those of the signals in figure 3.20. Indirect measurements are also, therefore, the most effective approach in this case. In this way, the variation in the delay caused by coupling can be evaluated by looking into the effect that this variation has on a given circuit.

The circuit considered for the measurements is a flip-flop one. In any synchronous circuit, there is a direct relationship between the minimum clock period allowed and the logic path delay:

$$T_C > D_Q + D_{LM} + U \tag{3.22}$$

being T_C the clock period, D_Q the flip-flop delay, U the setup time, and D_{LM} the maximum logic path delay. If one of the nodes in the logic path is coupled to another line of the circuit and there is a simultaneous transition in that second line, the total path delay of the synchronous circuit will be affected. Two different cases can be distinguished:

1 The transition in the affecting line is in the same direction as the transition in the affected line. In this situation, the affecting signal causes a decreased transition delay and the circuit may operate correctly for a value of clock period smaller than that allowed when there is no crosstalk between these signals.

2 The transition in the affecting line is in the opposite direction to that of the transition in the affected line. Now the affecting signal causes an increased transition delay and, in this case, if the clock period is near the minimum allowed, faulty behavior may be observed due to violation of equation 3.22.

Thus, by measuring the minimum allowed clock period (following relation 3.22) of a synchronous circuit with and without crosstalk coupling, it is possible to estimate the coupling effect introduced to the delay.

In order to show the applicability of this technique to measure crosstalk- induced delay, a specific integrated circuit has been designed and manufactured. The circuit diagram can be seen in figure 3.21 while the photograph of the chip is shown in figure 3.22. The circuit consists of a D-type flip-flop, with its output inverted by a logic block, which feeds the D input of the flip-flop. In this way, as the D input is the flip-flop output inverted and delayed, the normal operation of the circuit is a periodic oscillation with frequency half the clock frequency. The inverting logic block consists of one NAND gate with an external input which enables or disables the oscillation, plus a group of four inverters, then a long metal 2 line (victim) running in parallel to a second line (affecting line) separated by the minimum distance allowed by the technology and driven by another external input, and then a second group of six inverters. The length of the coupled lines is 1 mm approximately. The logic path presents a delay of 20 ns approximately (according to HSPICE simulations). The layout parameter extraction gives a capacitance to ground of 198 fF for the victim line, 175 fF for the affecting one, and a coupling capacitance of 46 fF between both.

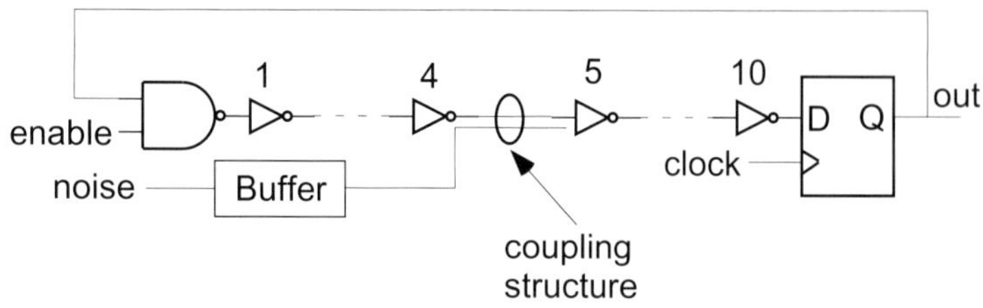

Figure 3.21. Schematic of the circuit considered for crosstalk delay evaluation.

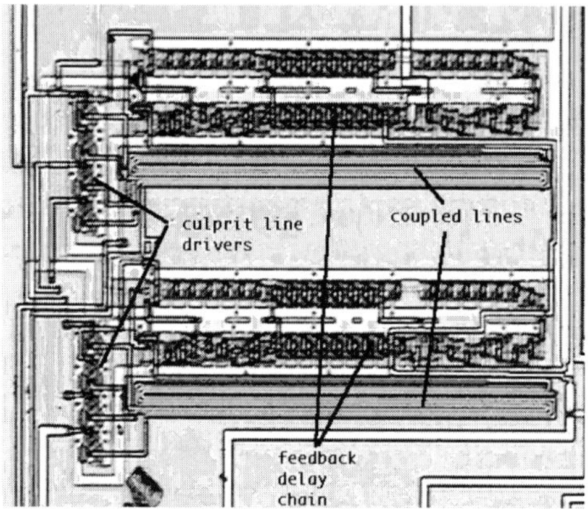

Figure 3.22. Photograph of the circuit considered for crosstalk delay evaluation.

When coupling introduces changes in propagation delay, the different behaviors that can appear may be classified into the following types:

- Introduce small increase in delay.
- Introduce a great increase in delay.
- Decrease in propagation delay.

Figures 3.23 and 3.24 illustrate the expected faulty output for a small and great induced delay. In figure 3.23, a comparison of the different behavior of the circuit without crosstalk and with crosstalk is shown. At the top of the diagram, the minimum period clock waveform allowing correct behavior is represented. It is assumed that the delay between the flip-flop output and the D input of the flip-flop is 5 time-units (the same as for HL and LH transitions). The set-up time of the flip-flop is 2 time-units, and the time delay of the flip-flop is 1 time-unit. As can be seen at the top of the diagram, the flip-flop input changes just before 2 time-units of the positive edge of the clock signal, and then, after 1 time-unit, the flip-flop output takes the value of its input. In this situation, the correct behavior is a signal of twice the period of the clock at the flip-flop output.

At the bottom of this diagram (separated from the top by the axis representing the time-units considered) the clock signal, flip-flop input and flip-flop output are shown. Now, due to noise in the feedback of the signal, the delay between the output of the flip-flop and input of the same module changes to a value of 6 time-units for the LH transition at the input and 5 time-units (as in the correct behavior) for the HL transition on D. Therefore, a violation of the flip-flop set-up time appears in the first case and, as seen in the diagram, the flip-flop output cannot follow all the changes to the input. In this case, the behavior of the circuit changes, and an oscillation is observed with a period three times the period of the clock signal at the output. We call this situation a case of minor delay, because violations are only produced at one of the edges.

Figure 3.24 analyzes the case known as a major delay, where violations of flip-flop set-up time are produced in both edges of the signal. At the top of this diagram, correct behavior is again represented so it can be compared with the faulty behavior. At the bottom of the diagram 3.24, the delay time between flip-flop output and input is now 6 time-units for both transitions (HL and LH). A set-up violation is produced for both edges and the flip-flop output now shows oscillating behavior with four times the period of the clock signal.

Finally, if the noise means a decrease in the delay of the signal between flip-flop output and input, the result will be that the clock period can be reduced from the "normal" minimum (without crosstalk-induced delay), and the correct behavior (flip-flop output oscillating at twice the period of the clock signal) will still be shown. As a numerical example, following the same data as the previous examples, if the delay is now reduced by 1 time unit, that means the delay between flip-flop output and input is now 4 time units for both signal edges, then the clock period (which is 8 time units in the previous examples) can be reduced to 6 time units and correct behavior is maintained.

Given the above analysis, the measurement procedure will be firstly to determine the minimum clock period without crosstalk (T_{mn}) (no activity is forced in the affecting line) that gives a correct output. After that, taking a clock period smaller than the previously measured minimum, a periodic signal with half the clock frequency is applied to the affecting line, such that it reduces the transition delay in the victim line and the circuit still behaves correctly. By measuring the minimum clock period under these conditions, (T_{md}), it is possible to determine the delay subtracted by coupling, which will be given by the difference between both minimum clock periods.

In contrast, to determine the added delay due to crosstalk, one must first start with a clock period greater than the normal minimum, T_{mn}. Then, forcing a signal in the affecting line so that the path delay is increased, if faulty behavior is observed it means that the initial clock frequency is smaller than

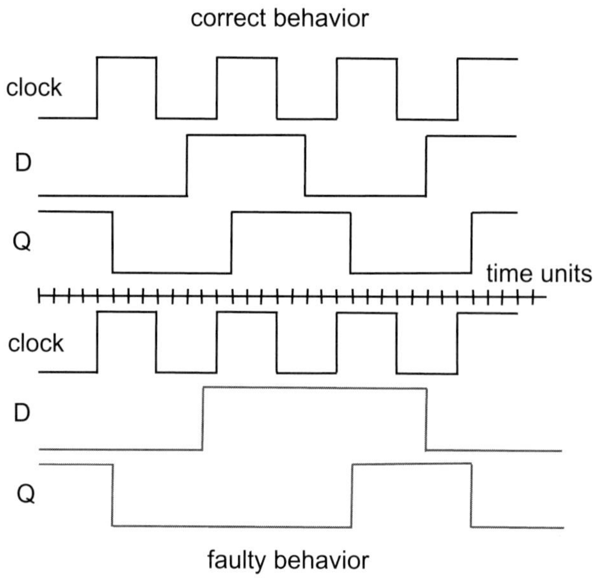

Figure 3.23. Comparison between correct behavior and minor delay behavior.

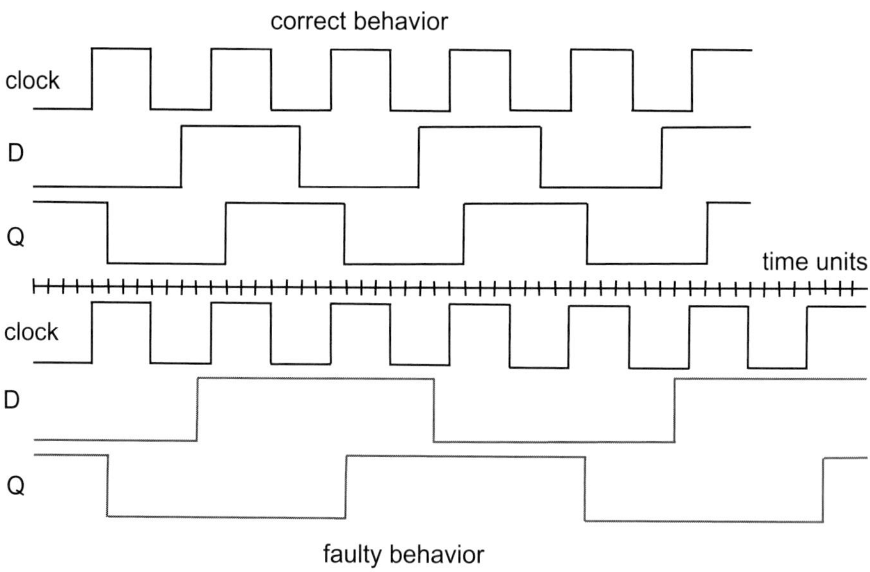

Figure 3.24. Comparison between correct behavior and major delay behavior.

the minimum with added delay, T_{ma}. This T_{ma} can be found by consecutively increasing the initial clock period until no faulty behavior is observed.

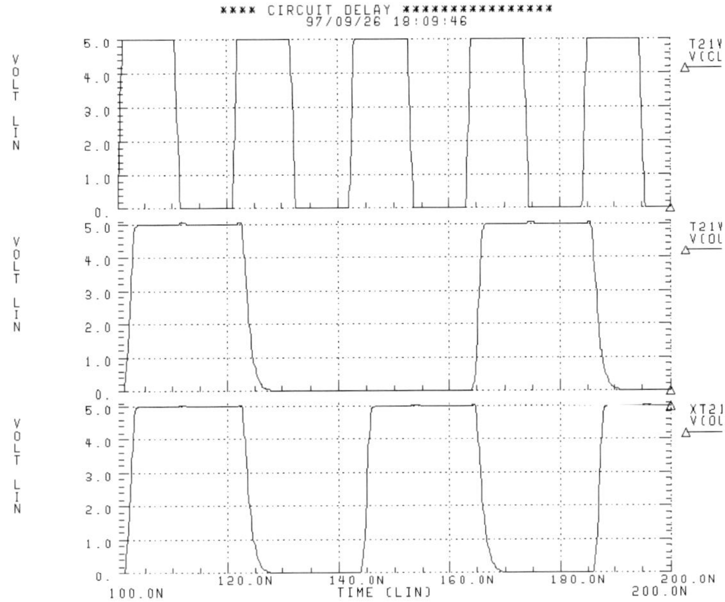

Figure 3.25. Simulation results. Clock signal (top), circuit output without crosstalk (middle) and circuit output with crosstalk (bottom)

To illustrate the method, figure 3.25 shows some simulation results using parameters extracted from the layout and device models provided by the foundry. At the top, the clock signal (21 ns period) can be seen. The middle waveform represents the circuit output without crosstalk, showing faulty behavior: a waveform with a period three times the clock period. This means that the normal operation of the circuit does not allow a clock period of 21 ns, and a slower clock is needed (in this case the minimum clock period determined by simulation is 22 ns). The waveform at the bottom represents the output of the circuit with crosstalk (implying a reduction in signal delay), and, as can be seen the circuit can operate correctly with a clock period of 21 ns. Using this procedure in the experiment, an evaluation of the delay introduced by coupling can be made.

2.2 Experimental results

The chip on which the measurements were performed has two identical sets of six lines, differing only in the victim line: in one of them the victim line is laid out in Metal 1, while in the other the victim line is in Metal 2, as can be seen schematically in figure 3.26. Each set will be measured separately. For each set of measurements, we can distinguish the following cases:

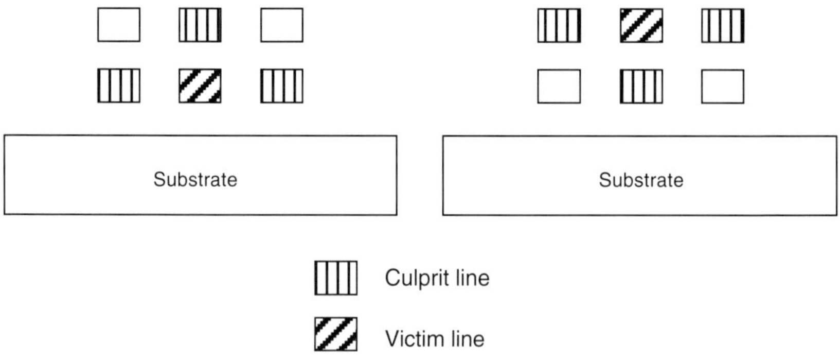

Figure 3.26. Schematic cross section of the coupled lines in the circuit photographed in figure 3.22.

Case 1 only one laterally adjacent line is switching simultaneously with the victim.

Case 2 two adjacent lines are simultaneously switching.

Case 3 the vertically adjacent line is causing the crosstalk.

Case 4 three adjacent lines switch simultaneously with the victim.

As the nature of the errors is somewhat random, especially at the transition between normal and faulty behavior, a statistical measurement based on error rate is shown. The oscilloscope is set up to be triggered by pulses separated by more than 1.5 times the clock period. First we start by choosing the minimum clock period without crosstalk. Figure 3.27 shows the measurement performed on the circuit applying a clock period of 23.95 ns (situation where victim is in Metal 2). As can be seen, the circuit shows a correct output for several clock cycles, but a faulty output (denoted by output pulse width greater than the clock period) appears at certain time intervals. By making a number (500) of requests asking for the triggering status of the oscilloscope, it is possible to determine a fault percentage, which is the magnitude displayed in the vertical axis of figures 3.28 and 3.29. In these figures, the four measured coupling cases are shown. For each case, two sub-cases are depicted labeled I and D. The I cases correspond to the situation in which crosstalk noise introduces an increase in the path delay, while the D cases correspond to a decrease in the path delay, allowing a higher working frequency.

Figure 3.28 shows that, for the set with victim line in Metal 2, the minimum clock period allowing normal operation is approximately 23.98 ns when there is no coupling. If a crosstalk signal is present and it is properly synchronized with the internal victim signal, the minimum clock period can be as low as

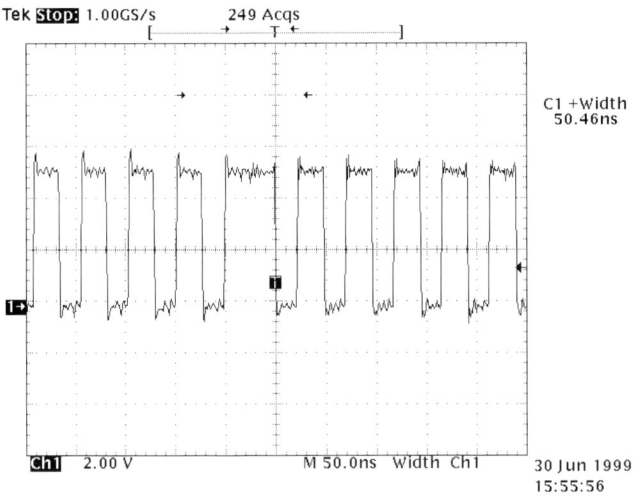

Figure 3.27. Measured waveform at the circuit output.

23.40 ns, which means a crosstalk effect of around 0.6 ns. This situation corresponds to Case 4, where three adjacent lines simultaneously interfere with the victim line.

The same measurements for the situation where the victim is in Metal 1 level are shown in figure 3.29. In this case, the same trends as in figure 3.28 are observed, but the crosstalk delay effect is less significant than when Metal 2 level is the victim. In the previous case, differences in delay of 600 ps were obtained, while now these differences are around 350 ps in the worst case (Case 4). This difference can be explained considering that Metal 1 presents a capacitance to the substrate that decreases the crosstalk effect.

3. Energy dissipation due to crosstalk

One important problem in today's VLSI integrated circuits is the evaluation of circuit energy consumption. The reason is the need to know with a good degree of accuracy the power consumption (the energy of one operation divided by the time this operation takes to execute) of the chip in order to consider heat dissipators properly, to design power supply and ground lines adequately (with enough width to avoid electro-migration problems) and to increase power autonomy in portable electronic equipment.

Technological trends dictate an increase in capacitance between lines [14]. In the usual analysis of switching power consumption, this capacitance is ignored, or at least, included implicitly in a total capacitance to ground [15].

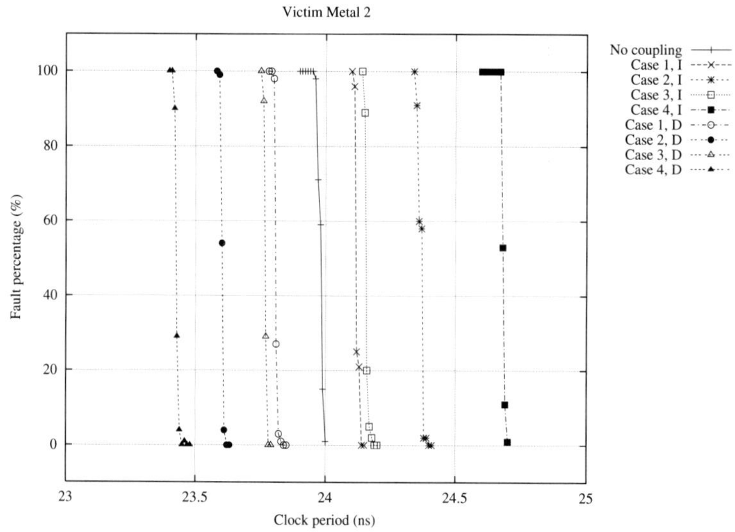

Figure 3.28. Measurement results for the situation when victim in metal 2 level.

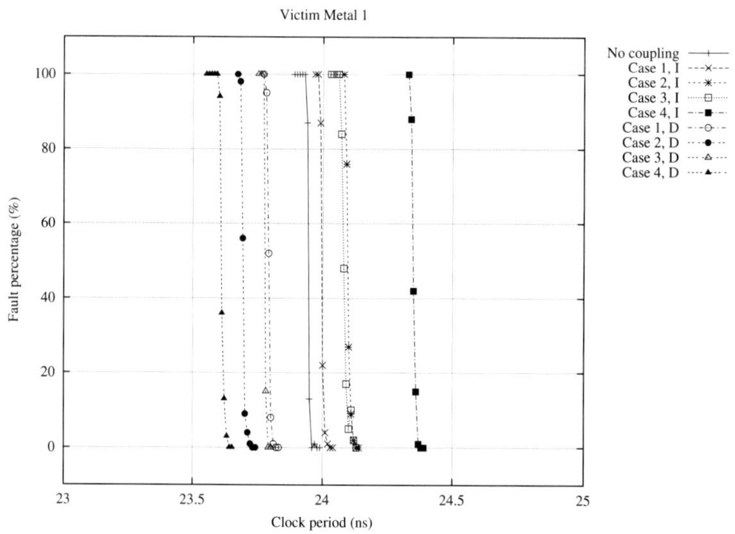

Figure 3.29. Measurement results for the situation when victim in metal 1 level.

However, the effect of this coupling capacitance is very different from capacitance to ground because it depends on the switching activity of the drivers. An estimate in [16] gives an increase factor of 4 in energy consumption due to coupling between lines.

The increase in energy dissipation due to interconnect structures shows two different contributions: the first one is the spurious signal produced because of the coupling capacitance, which causes extra dissipation in the coupled lines' driver resistance. The second contribution is due to the logical propagation of this spurious signal in the nodes following the coupled signal. The first contribution will be quantified in this section, showing how much more energy is consumed by the presence of capacitance between lines. This aspect was mentioned in [17]. An analysis of the energy consumption considering transistor devices, formally including short-circuit, sub-threshold and input-output capacitance current contributions, is developed in this section.

A set of different line structures is analyzed through electrical simulation in order to calculate the energy dissipated, taking into account the lines' coupling capacitance. An evaluation of the different contributions to the total energy dissipated will be made, allowing the introduction of some rules depending on the structure, which are very useful in energy estimation tools for optimizing the algorithms they use.

We will start by presenting a simple model for a set of two coupled lines with CMOS inverters as drivers. Single transitions will be analyzed from a mathematical point of view. Some simulations will be also performed and the relative significance of different contributions for total energy dissipation will be analyzed. This section will end with the case of multiple transitions.

3.1 Model for energy calculation of two coupled lines. Case of single transition.

In order to calculate the dynamic current consumption of two coupled lines, let us consider the circuit depicted in figure 3.30. Two interconnect lines are driven respectively by two CMOS inverters. Each line is described as a capacitance to ground and a coupling capacitance between them, modeling capacitive crosstalk. For the electrical structure of figure 3.30, the relation between the different currents is given by the following expressions:

$$i_{p1} - i_{n1} = (C_1 + C_{12} + C_{c1})\frac{dV_1}{dt} - C_{c1}\frac{dV_{i1}}{dt} - C_{12}\frac{dV_2}{dt} \tag{3.23}$$

$$i_{p2} - i_{n2} = (C_2 + C_{12} + C_{c2})\frac{dV_2}{dt} - C_{c2}\frac{dV_{i2}}{dt} - C_{12}\frac{dV_1}{dt} \tag{3.24}$$

The energy dissipated in the transistors during a certain time interval can be obtained by integrating the power, that is the voltage drop in each transistor multiplied by the current:

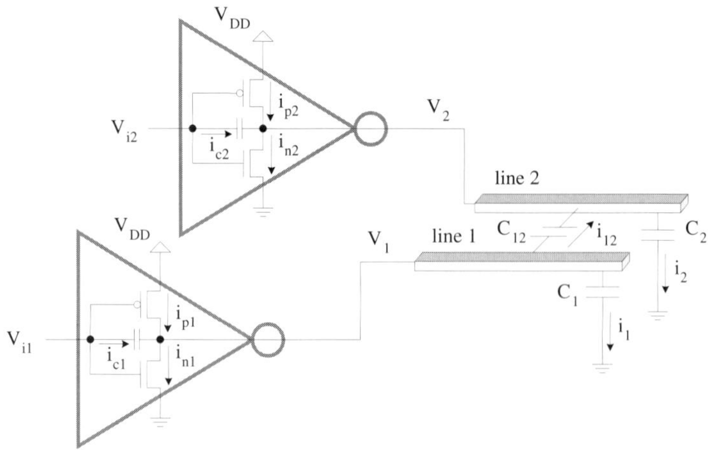

Figure 3.30. Electrical circuit modeling the structure of two coupled lines with inverter drivers.

$$
\begin{aligned}
E_T &= V_{DD}\int i_{p1}dt + V_{DD}\int i_{p2}dt - \int V_1\left(i_{p1}-i_{n1}\right)dt - \\
&\quad \int V_2\left(i_{p2}-i_{n2}\right)dt \qquad (3.25)
\end{aligned}
$$

Let us first analyze the case of a positive transition in V_{i1}, while the driver input of the other node (V_{i2}) is kept at a constant value, V_{DD}. In this case, V_{i2} is constant and therefore its derivative and increment are null. Driver 1 is making a positive transition and thus the current through transistor P_1 is the driver's short-circuit current contribution. On the other hand, driver 2 has a constant value V_{DD} at its input, and, therefore, the current through transistor P_2 is only a sub-threshold contribution. Substituting equations 3.23 and 3.24 into 3.25 gives:

$$
\begin{aligned}
E_T &= V_{DD}\int i_{p1,SC}dt + V_{DD}\int i_{p2,ST}dt - \\
&\quad \left(C_1+C_{12}+C_{c1}\right)\frac{1}{2}\Delta\left(V_1^2\right) + C_{c1}\int V_1\frac{dV_{i1}}{dt}dt - \\
&\quad \left(C_2+C_{12}+C_{c2}\right)\frac{1}{2}\Delta\left(V_2^2\right) + C_{12}\Delta\left(V_1V_2\right) \qquad (3.26)
\end{aligned}
$$

A spurious signal due to crosstalk will be produced at node 2, but after the steady state is reached, the final output of node 2 will be the same as at the beginning, hence $\Delta V_2 = 0$. Then, taking a sufficiently long integration period, the output voltages will reach stable values, and the resulting voltage differences will be $\Delta V_1 = -V_{DD}$, $\Delta(V_1^2) = -V_{DD}^2$, $\Delta(V_1 V_2) = 0$ and $\Delta V_2 = \Delta V_2^2 = 0$. Inserting these values in equation 3.26 leads to the following expression:

$$\begin{aligned} E_T &= V_{DD} \int i_{p1,SC} dt + V_{DD} \int i_{p2,ST} dt + C_{c1} \int V_1 \frac{dV_{i1}}{dt} dt + \\ &+ \frac{1}{2} (C_1 + C_{12} + C_{c1}) V_{DD}^2 \end{aligned} \tag{3.27}$$

This result applies for a positive transition at the input of driver 1, while input of driver 2 is at V_{DD}. The other cases must also be studied: a positive transition in V_{i1} while V_{i2} is at GND, a negative transition in V_{i1} while V_{i2} is at GND and a negative transition in V_{i1} while V_{i2} is at V_{DD}. The methodology for analyzing these other cases consists of finding in equations 3.23 and 3.24, the switching currents, depending on voltage derivatives and non-switching currents: the current of the transistor in the sub-threshold region (cut-off from the quiet driver) and the current of the transistor which contributes with a short-circuit current (transistor of the active driver that is in cut-off state at the final stable state). Then, the energy is obtained from equation 3.25. In all cases, the total energy can be expressed as a sum of different contributions:

$$E_T = E_{SC} + E_{ST} + E_{IO} + E_{LINES} \tag{3.28}$$

In equation 3.28 the following contributions can be identified:

a) Short-circuit energy (E_{SC}) due to the transistor of the active driver (driver 1 in our case) which is cut off at the end of the transition. In the contribution of this transistor, a sub-threshold component is also present once the steady state is reached. This is included in the same i_{p1}.

b) Sub-threshold energy (E_{ST}) due to the quiet driver's (driver 2 in our case) transistor, which is cut off all the time.

c) A contribution at the active driver due to input-output parasitic capacitance of the devices (E_{IO}). This contribution is given by [18]:

- For positive transitions of V_{i1}:

$$E_{IO}^{(+)} = \frac{1}{2} C_{c1} V_{DD}^2 + C_{c1} \int V_1 \frac{dV_{i1}}{dt} dt \tag{3.29}$$

- For negative transitions of V_{i1}:

$$E_{IO}^{(-)} = \frac{3}{2} C_{c1} V_{DD}^2 + C_{c1} \int V_1 \frac{dV_{i1}}{dt} dt \tag{3.30}$$

d) A contribution due to interconnects (E_{LINES}). In all cases, given by

$$E_{LINES} = \frac{1}{2}(C_1 + C_{12})V_{DD}^2 \tag{3.31}$$

Contributions a), b) and c) depend on the driver's transistors, while contribution d) depends only on the interconnect structure.

A different problem appears if there are simultaneous transitions in both lines. In that case, a more complicated mathematical treatment is needed, and two new parameters play an important role: the relative direction of the transitions (opposite or equal direction) and, especially, the relative delay between transitions. A study of this problem will be made at the end of this section.

3.2 Contribution of driver and interconnect to dissipated energy. A case study

In order to analyze the importance of the different contributions to the energy consumption, some HSPICE simulations were conducted on the same circuit used to derive the analytical expressions in the previous section (figure 3.30). A 0.18 μm CMOS technology with one poly and six metallization levels is considered. HSPICE simulations have been performed using a level 49 model for the MOSFET devices, with a supply voltage of V_{DD} = 1.8 V. The energy dissipation values have been calculated by numerically integrating the simulated transient current through the transistors, multiplied by their voltage drop.

Parasitic capacitances (C_1, C_2 and C_{12}) have been calculated using the 2D field solver integrated into HSPICE [19]. Also, an estimated fan-out capacitance of $C_g = 5\,\text{fF}$ (equivalent to six minimum-size gates) has been added to the calculated capacitance to ground to give the node ground capacitances C_1 and C_2.

Four different structures have been considered in order to analyze the effect of different line geometries in the energy dissipation contribution of coupling (figure 3.31). For the coupling configuration labeled *Struct A*, different cases have been considered concerning the size of the driver transistors:

- Equal and unbalanced drivers (equally sized transistors $L = 0.18\,\mu\text{m}$, $W = 1\,\mu\text{m}$).
- Equal and balanced drivers ($L = 0.18\,\mu\text{m}$, $W_n = 1\,\mu\text{m}$, $W_p = 2.5\,W_n$).
- Different and balanced drivers ($L = 0.18\,\mu\text{m}$, $W_{n1} = 1\,\mu\text{m}$, $W_{p1} = 2.5\,W_{n1}$, $W_{n2} = 5\,W_{n1}$, $W_{p2} = 5\,W_{p1}$).

For the other coupling structures, only the case of equal drivers with equally sized transistors is reported. For all the cases considered, three energy values have been obtained:

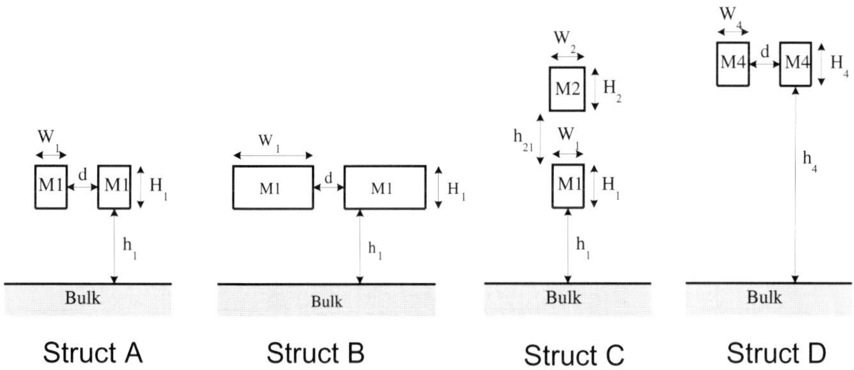

Figure 3.31. Structures considered in the analysis of coupling geometry influence on the energy consumption.

- The total energy including crosstalk effects (C_{12} value obtained from field solver), by HSPICE simulation.
- The total energy without considering crosstalk effects ($C_{12} = 0$), by HSPICE simulation.
- The line contribution to the total energy, labeled E_{LINES}, which can be analytically calculated.

Two comparisons are presented here. Firstly, the energy dissipation due to drivers (contributions a, b and c above) is compared with the energy dissipation due to interconnect capacitances (E_{LINES}). The latter component is obtained analytically, and the former component is obtained by subtracting the value of E_{LINES} from the total energy dissipation calculated by HSPICE. The second comparison is total energy obtained by HSPICE simulation, between the case with coupling capacitance, and with no coupling capacitance (i.e. repeating the simulation with $C_{12} = 0$).

Let us first focus on the geometry labeled *Struct A* in figure 3.31 that consists of two equal lines of minimum width and minimum separation at the first level of metallization. The results obtained are depicted in figure 3.32 for different line lengths. It can be seen that the drivers' contribution represents 50% of the total energy dissipation for a very short line (1 μm) and this decreases rapidly when line length increases. For example, for a line of 250 μm in length, the contribution of drivers is only around 10% of total energy consumption. This result shows that in energy dissipation analysis of VLSI circuits, the effect of lines can never be neglected, while the effect of drivers (due to short-circuit plus sub-threshold current) can be neglected for lines longer than a certain value. This result can be used in energy dissipation estimation tools to obtain

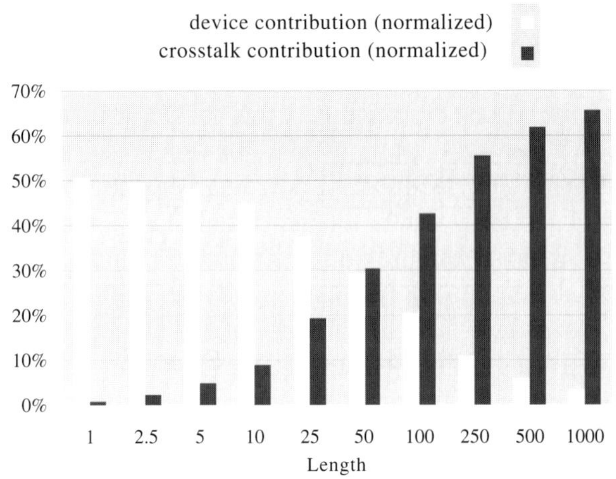

Figure 3.32. Drivers and crosstalk contributions to total energy consumption for structure A.

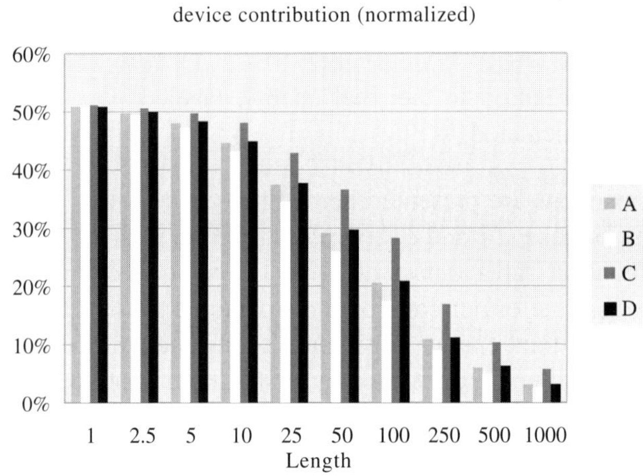

Figure 3.33. Drivers contribution to total energy dissipation, comparing A, B, C and D structures.

accurate results and improve analysis time. In figure 3.32 the contribution of crosstalk is also analyzed. It can be seen that for 100 μm long lines crosstalk contribution is around 40% of the total energy dissipation. Only for short lines can this contribution be neglected. This implies that energy estimation tools cannot disregard the contribution of crosstalk (coupling parasitic capacitance) because it represents a very important percentage of total energy dissipation.

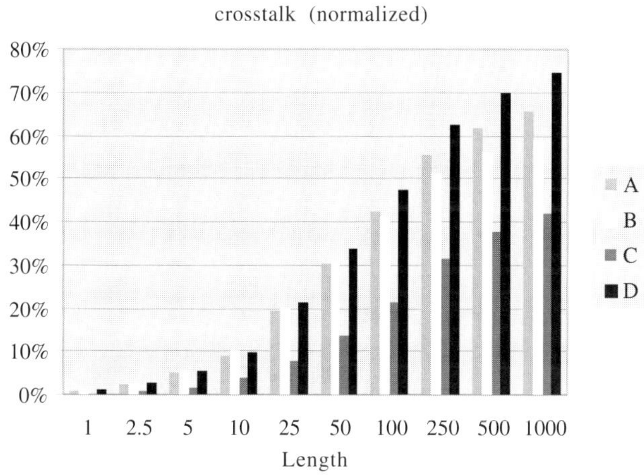

Figure 3.34. Crosstalk contributions to total energy dissipation, comparing A, B, C and D structures.

Figure 3.33 shows a comparison of drivers' contribution to energy dissipation for the four structures in figure 3.31 (A, B, C and D). The same trends can be observed in all structures. The device contribution is around 50% (equal contribution of drivers and lines) for short lines ($1\,\mu m$ length) and decreases for longer lines. For $100\,\mu m$ length lines, the drivers' contribution is around 20% of total energy dissipation, except for structure C, where this contribution is near 30%. It must be noted that for structure C, the drivers' contribution is more important than for the other structures, due to the screening effect of metal 1 on the upper metal layer, resulting in a small value for metal 2 ground capacitance, thus reducing the total capacitance involved in this case.

Figure 3.34 shows the importance of crosstalk contribution to total energy dissipation for the structures analyzed. Once more, similar trends are obtained in all the structures. For short lines, the effect of crosstalk is negligible, while for long ones it is very significant. For lines up to $100\,\mu m$ the crosstalk contribution represents more than 40% of total energy dissipation, except for structure C. Structure C shows less crosstalk significance (20% of total energy dissipation), for the above- mentioned reasons. In figure 3.34 it can also be seen that the crosstalk effect is more significant in structures A and D. This fact is related to the previously mentioned result that crosstalk noise increases when the ratio between coupling and ground capacitance increases. This is also true for extra energy dissipation. Comparing structures A and B, the coupling capacitance is almost the same, while the ground capacitance is greater in structure B due to the wider lines considered, reducing the ratio between coupling and ground capacitance. The same happens with structure D, but in

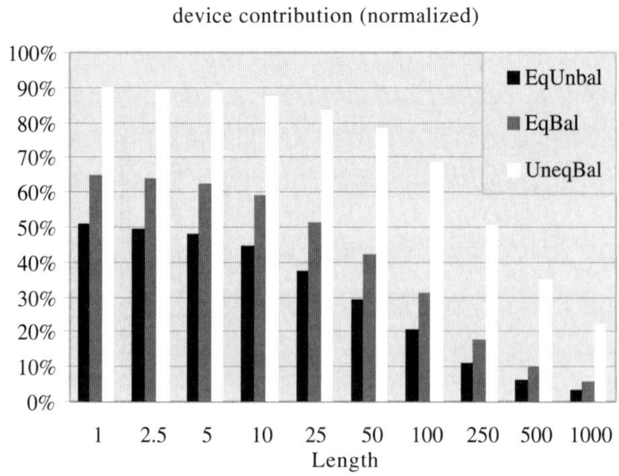

Figure 3.35. Driver dependence in energy dissipation of lines.

this case it is not so intuitive because of the different values for width, height and distance between lines.

Figure 3.35 analyzes the effect of driver sizes. It is observed that when line length increases, the drivers' relative contribution to total energy consumption decreases. Considering different driver sizes, it can be seen that the significance of the contribution of drivers is more relevant when the size of the drivers increases, as intuitively expected. Comparing the cases of unequal balanced and equal balanced drivers (the first case has transistors with a channel width five times that of the second case), the result is that, in unequal balanced case, the importance of drivers' contributions is more significant than for the equal balanced case (for a line length of $100\,\mu$m, the values change from 70% to 30% respectively).

3.3 Transitions in both nodes

In this section, the more complicated case of non-simultaneous transitions at sources is dealt with. The total dissipated energy calculation will be split in two phases, A and B (see figure 3.36), so that each phase corresponds to a single transition. The main difference with respect to the single transition is that phase A has an integration period which is not infinite, as in the case of the energy computation for a single switching node, but lasts only until the second transition (phase B) begins. Thus, at the end of phase A, $\Delta V_2 = D_{2A} \neq 0$ and $\Delta V_1 = D_{1A} \neq \pm V_{DD}$. Voltage differences D_{1A} and D_{2A} are explicitly shown in figure 3.36.

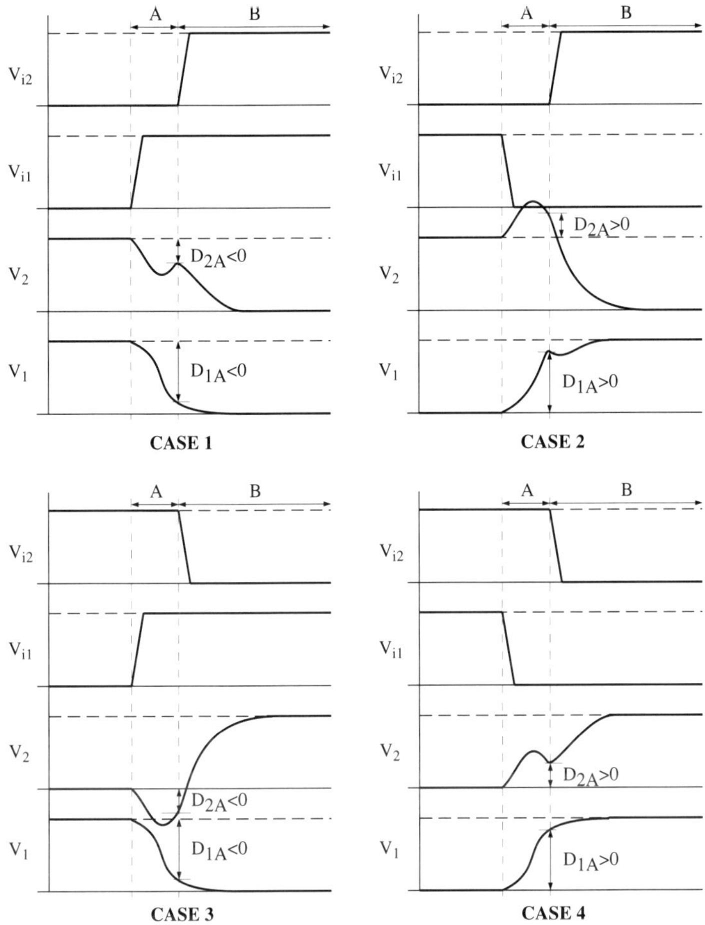

Figure 3.36. Schematic transient waveforms for two coupled outputs. Phases A and B for the total energy calculation in the case of non- simultaneous transitions. All 4 possible cases are shown.

Following the notation in figure 3.36 and after some algebra applying equations for a single transition in all the cases (different direction transitions and different values at quiet nodes), it is possible to calculate the total energy dissipated in terms of these intermediate values for the different switching event cases. A detailed analysis can be found in [18].

As a summary of the analysis, there are only two different cases to consider: transitions in the same direction and transitions in opposite directions. In terms of the *absolute value* of D_{1A} and D_{2A}, we have for opposite transitions:

$$E_T = E_{SC} + E_{ST} + E_{IO} + \frac{1}{2}(C_1 + C_{12})V_{DD}^2 + \frac{1}{2}(C_2 + C_{12})V_{DD}^2 + \Delta \quad (3.32)$$

where Δ depends on D_{1A} and D_{2A} as well as on the coupling capacitance, and can be expressed as:

$$\Delta = C_{12}V_{DD}^2 + C_2 V_{DD}|D_{2A}|\left[1 + \frac{C_{12}}{C_2}\left(1 - \frac{|D_{1A}|}{|D_{2A}|}\right)\right] \quad (3.33)$$

On the other hand, for transitions in the same direction:

$$E_T = E_{SC} + E_{ST} + E_{IO} + \frac{1}{2}(C_1 + C_{12})V_{DD}^2 + \frac{1}{2}(C_2 + C_{12})V_{DD}^2 - \Delta \quad (3.34)$$

In the previous equations, E_{SC} and E_{ST} correspond to the transistors' subthreshold and short-circuit current as before, while E_{IO} corresponds to all terms proportional to C_{c1} or C_{c2}.

3.4 Discussion

Assuming the usual case is that the maximum occurs at $D_{1A} = D_{2A} = 0$, for transitions at both nodes in opposite directions (Cases 2 and 3), the effect is an increase in energy. This maximum effect can be calculated, giving for both cases:

$$E_{T,\max} = E_{SC} + E_{ST} + E_{IO} + \frac{1}{2}C_1 V_{DD}^2 + \frac{1}{2}C_2 V_{DD}^2 + +2C_{12}V_{DD}^2 \quad (3.35)$$

In the case of signals switching the same direction (Cases 1 and 4), a decrease in energy is obtained:

$$E_{T,\min} = E_{SC} + E_{ST} + E_{IO} + \frac{1}{2}C_1 V_{DD}^2 + \frac{1}{2}C_2 V_{DD}^2 \quad (3.36)$$

This is the base for the algorithms which attempt to find codes minimizing the number of opposite transitions in coupled lines, and favoring transitions in the same direction [20], [21], [22]. However, if one of the transitions is delayed with respect to the other, at least one of D_{1A} or D_{2A} will be non-zero and the energy dissipated will have a value between the case of two independent transitions, and the above $E_{T,\min}$ or $E_{T,\max}$. Therefore, these algorithms will be useful when "simultaneous" transitions are really simultaneous and are not skewed by different logic paths by an amount greater than a certain limit.

3.4.1 Effect of relative delay between input transitions

These effects will be illustrated with several HSPICE simulations. We have used a level 49 MOSFET model (BSIM3v3) of a 0.18μm CMOS technology. Power supply is chosen to be 3 V, and input signal rise time, 200 ps. The loading capacitances are C_1=400 fF, C_2=300 fF, C_{12}=250 fF. Three configurations are studied:

1 Both drivers are equal, with equally- sized PMOS and NMOS transistors, L=0.34μm, W=2μm (unbalanced, equal drivers).

2 Drivers are equal and balanced, L=0.34μm, W_n=2μm, W_p=5μm.

3 Drivers are different and balanced: L=0.34μm, W_{n1}=1μm, W_{p1}=2.5μm, W_{n2}=2μm, W_{p2}=5μm.

We calculate the energy dissipation obtained from expressions 3.35 and 3.36 without including technological contributions ($E_{ST} = 0$, $E_{SC} = 0$, $E_{IO} = 0$ and $C_{cx} = 0$), as they are assumed much smaller than loading capacitance contribution for long lines. The values thus obtained are 2.925 pJ for a single transition at node 1, 2.475 pJ for a single transition at node 2, 3.15 pJ for simultaneous transitions in same direction and 7.65 pJ for opposite direction simultaneous switching. These values compare well with those in table 3.1 for the configurations mentioned above. After careful evaluation of the simulation results, the differences between the three cases shown in table 3.1 can, in this case, be attributed mainly to various contributions in short-circuit current and I/O coupling capacitance. An overall contribution of the short-circuit current, sub-threshold current, and I/O coupling capacitance is obtained of between 50 fJ and 75 fJ per driver, given this particular configuration.

With these values, the energy obtained from HSPICE as a function of relative delay between input transitions is shown in figures 3.37 to 3.39. These diagrams show that the influence of coupling capacitance in the case of double transitions is only important for a certain range of relative delays. The question of what this particular range is therefore arises. From the analysis in the preceding sections, it is clear that this range corresponds to the situation in which $|D_{1A}|$ (the output variation in the first switching output) is less than V_{DD}, or $|D_{2A}|$ (the output variation in the coupled line due to the first switching output) is non-zero. In other words, the range in which crosstalk coupling has a

Output Transition	*Unbalanced Equal*	*Balanced Equal*	*Balanced Unequal*	*Theor. Switch.*
11 → 00	3.245	3.324	3.282	3.150
00 → 11	3.242	3.299	3.263	3.150
10 → 01	7.744	7.814	7.770	7.650
01 → 10	7.744	7.813	7.776	7.650
11 → 10	2.522	2.563	2.563	2.475
10 → 11	2.521	2.550	2.550	2.475
11 → 01	2.972	3.013	2.970	2.925
01 → 11	2.972	3.000	2.963	2.925

Table 3.1. Dissipated energy from HSPICE simulations, in pJ, for coincident transitions, and single transitions. The last column shows theoretical energy, considering only energy due to switching output capacitances.

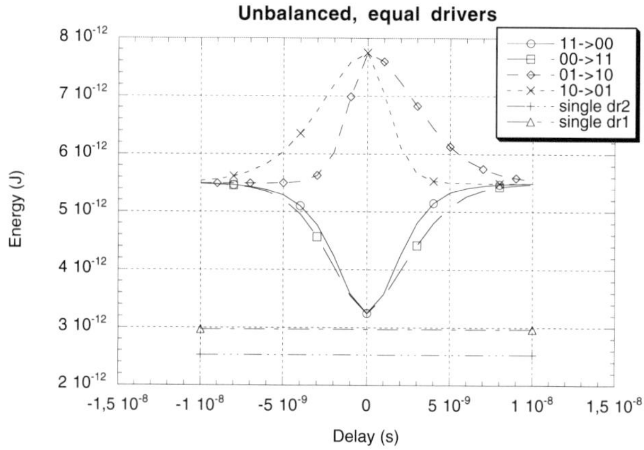

Figure 3.37. Energy dissipation vs. relative delay between input transitions for unbalanced, equal drivers.

significant influence is determined by the slowest response of either the first transition ($|D_{1A}|$) or the corresponding crosstalk waveform ($|D_{2A}|$).

Observe, for example, figure 3.37. As both drivers are unbalanced (PMOS response is slower than NMOS's), the range of relative delay in which coupling capacitance is significant is greater for $0 \rightarrow 1$ output transitions than for $1 \rightarrow 0$ transitions. A negative delay indicates that the transition at the input of driver 2 precedes that of driver 1, so for negative delay, transition $10 \rightarrow 01$ ($0 \rightarrow 1$ in line 2) has a greater range of influence than $01 \rightarrow 10$ transitions, while the opposite occurs for positive relative delays. This is shown in the clear asymmetry of the curves $01 \rightarrow 10$ and $10 \rightarrow 01$ with respect to the origin in

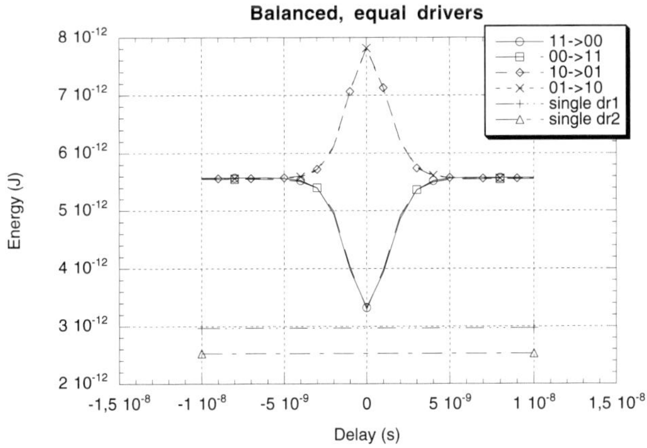

Figure 3.38. Energy dissipation vs. relative delay between input transitions for balanced, equal drivers.

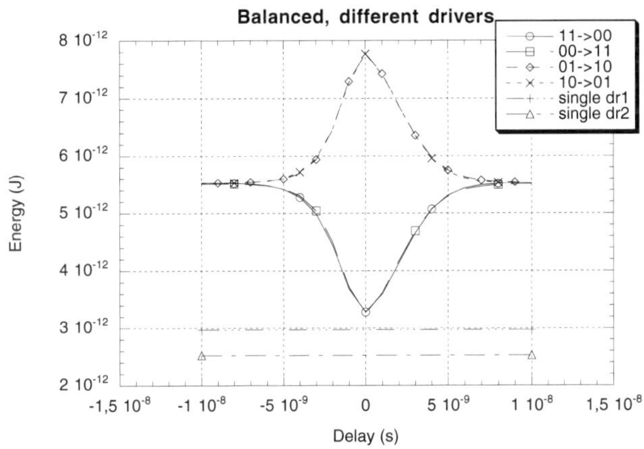

Figure 3.39. Energy dissipation vs. relative delay between input transitions for balanced, different drivers.

the x-axis. Also, the difference between curves $00 \rightarrow 11$ and $11 \rightarrow 00$ is due to the same cause.

This asymmetry does not appear in figure 3.38, as the response of PMOS and NMOS is approximately the same and drivers are equal. As for figure 3.39, there is an asymmetry coming from the different strength of each driver: driver 2 is stronger and has a faster response (equal for $0 \rightarrow 1$ or $1 \rightarrow 0$ as the drivers are balanced). Therefore, when there is a positive relative delay (driver 1 switches first), the coupling capacitance influence has a greater extent than for negative relative delays, as the figure shows.

In general, as a rule of thumb, we can conclude that the range of influence of coupling capacitance is directly related to the output response, or output rise time. Input transitions delayed by more than the value of the output rise time will not produce energy dissipation significantly different to that of two separate transitions.

3.4.2 Influence of relative driver strength

The formulation presented here allows an explanation of the dependence of energy consumption on the characteristics of the output. Let us consider the output response of two equal drivers to a double transition delayed by a certain amount of time. If we compare the result to the response of two drivers in which one of them (namely, gate 1, the first switching) is more resistive, we will observe that, in the latter case (driver 1 more resistive), the corresponding voltage difference at the end of Phase A, $|D_{1A}|$ is smaller than in the former case (equal drivers). Also, $|D_{2A}|$, the voltage difference in the victim node, will be quite similar because driver 2 has the same strength in both cases. Observing equations 3.32 and 3.34, we can conclude from this rough analysis that in the second case the energy dissipation will be larger if the transitions are in opposite directions, and smaller if the transitions are in the same direction. In general, then, we can conclude that if driver 1 is more resistive (weaker) than driver 2, opposite transitions tend to worsen the energy dissipation and transitions in the same direction alleviate it, compared to the case of equal driver strengths. The opposite occurs when driver 1 is stronger than driver 2.

Figure 3.40 shows the influence of relative driver strength on output waveforms. As driver 1 strength decreases with respect to driver 2, the value of $|D_{1A}|$ also decreases, while $|D_{2A}|$ hardly varies. Therefore, as predicted by equation 3.34, the dissipated energy must increase as driver strength increases for this particular case (both transitions in the same direction). This is confirmed by the results obtained from HSPICE simulations, shown in table 3.2. The only exception is case 16/4 which for opposite transitions should show lower energy than 8/4 case according to the rule of thumb we use. In this case, as can be seen from the waveforms in figure 3.40, driver 1 is so fast that the transitions are essentially independent of a relative input delay of 1 ns, and the resulting energy is approximately the sum of energy for each transition. Note that the increase in transistor size of driver 1 implies an increase in E_{SC}, E_{ST} and E_{IO} with respect to the theoretical values obtained only from line parameters shown in table 3.1.

4. Crosstalk effects in logic VLSI circuits

In this section we will discuss how crosstalk effects are translated into logic effects in digital circuits. This section is divided in two sections: one for static circuits, and the other for dynamic circuits.

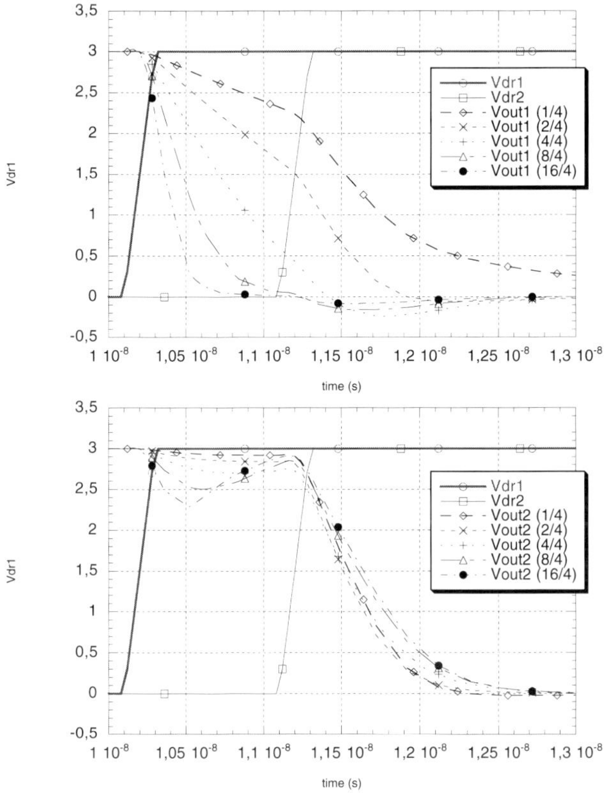

Figure 3.40. Output waveforms of gate 1 (top) and gate 2 (bottom) for different ratios between driver strength. Labels refer to the value of NMOS width, in μm, for the respective drivers. PMOS width is, in all cases, 2.5 times the corresponding NMOS width.

Transition	*1/4*	*2/4*	*4/4*	*8/4*	*16/4*
11 → 00	3.80	4.25	4.97	5.68	6.26
00 → 11	3.82	4.29	5.01	5.63	6.19
10 → 01	7.42	7.04	6.42	6.16	6.49
01 → 10	7.44	7.03	6.42	6.17	6.26

Table 3.2. Dissipated energy from HSPICE simulations, in pJ, for different relative driver strengths and combinations of output transitions. The input transition of driver 2 is delayed by 1 ns with respect to input transition of driver 1 (line 1 switches first).

4.1 Static circuits

As already pointed out, there are two effects that can cause problems in combinational blocks. One is an increase or decrease in the delay of the circuit. The difficulty in handling this extra delay is that it is data-dependent, because otherwise it could just be added to the delay associated with devices and routing. This data-dependency may appear as random errors produced by set-up or hold time violations when the delayed transition arrives at a storage element (a latch or a flip-flop).

The second effect is the generation of spurious signals. They introduce only a transient error in the circuit. The problem, again, is that this transient error may reach a storage element and then a wrong logic value may be stored. The study of spurious signal propagation must determine if the penetration depth is greater than the distance (measured in number of gates traversed) to a storage element, or to a primary output, where it can affect other chips. The propagation issue, besides determining the penetration depth in terms of the signal's amplitude and width, must consider the set of logic values that allow propagation through a logic path[1]. This latter aspect is dealt with in chapter 6.

In summary, the problem with crosstalk in static circuits is the possibility of storing a wrong logic value, either by an induced (positive or negative) delay, or by a spurious signal propagated to the input of a storage element. In addition, there is increased energy dissipation, as analyzed above.

4.2 Dynamic circuits

Dynamic circuits are commonly used in high-speed design because they show attractive features which include reduced area (the number of transistors usually is much smaller than the number of transistors in typical CMOS complementary technologies), and increased speed. However, this logic style is more susceptible to coupling noise because in the normal operation mode, during some time periods some nodes of the gate are in a high impedance state. As already shown, during these time intervals these internal nodes are therefore more susceptible to crosstalk noise. In this section we will show how crosstalk noise can affect this kind of circuit through two examples of dynamic circuits: a full adder as a combination circuit, and a dynamic RAM cell as a storage element.

4.2.1 NORA full adder

Let us start by considering a serial full adder implemented using the NORA technique [23]. Figure 3.41 shows the transistor diagram of the circuit. Several nodes of this structure, such as nodes labeled $Sum\_$, $Cout\_$, *Sum* and *Cout* are good candidates to be highly crosstalk sensitive as they are in a high impedance state during some semi-periods of the clock input (Φ). As an example, the

[1] For example, a spurious signal at one input of a 2-input NAND gate can never be propagated to its output if the other input is set to logic value 0.

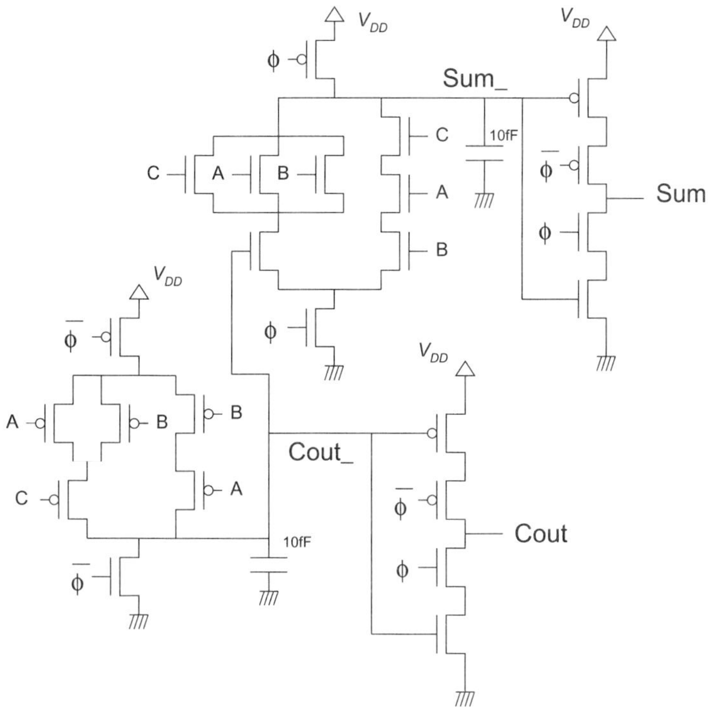

Figure 3.41. Electrical schema of a serial full adder implemented using NORA techniques.

effects of the simultaneous transitions of 3 lines which are coupled by parasitic capacitances to line $Cout\_$ have been simulated. Device models for a CMOS 0.35 μm and 3.3 V voltage supply technology are used in the circuit simulation. The values of parasitic capacitance between lines are of the same order than the ground capacitances of the lines.

Figure 3.42 presents the HSPICE simulation results of the full adder for several input values. Panel (c) illustrates how the transitions of the affecting lines affect the victim line $Cout\_$. A spurious signal can be observed of around 1.8 V in amplitude and less than 1 ns duration in this signal. Panel (d) shows the outputs *Cout* and *Sum* of the full adder without coupling, that is, in normal operation mode. Panel (e) shows the same result considering coupling in $Cout\_$. A permanent logic error in this output is observed in this panel. These errors can be further propagated through other logic blocks fed by the adder outputs.

From the point of view of circuit type, a dynamic full adder like the one presented here is a combinational circuit, but, in order to work properly, it needs to include some latch structures to maintain values at certain nodes. In essence, this is what allows the increase in speed of dynamic circuits: in addition to a

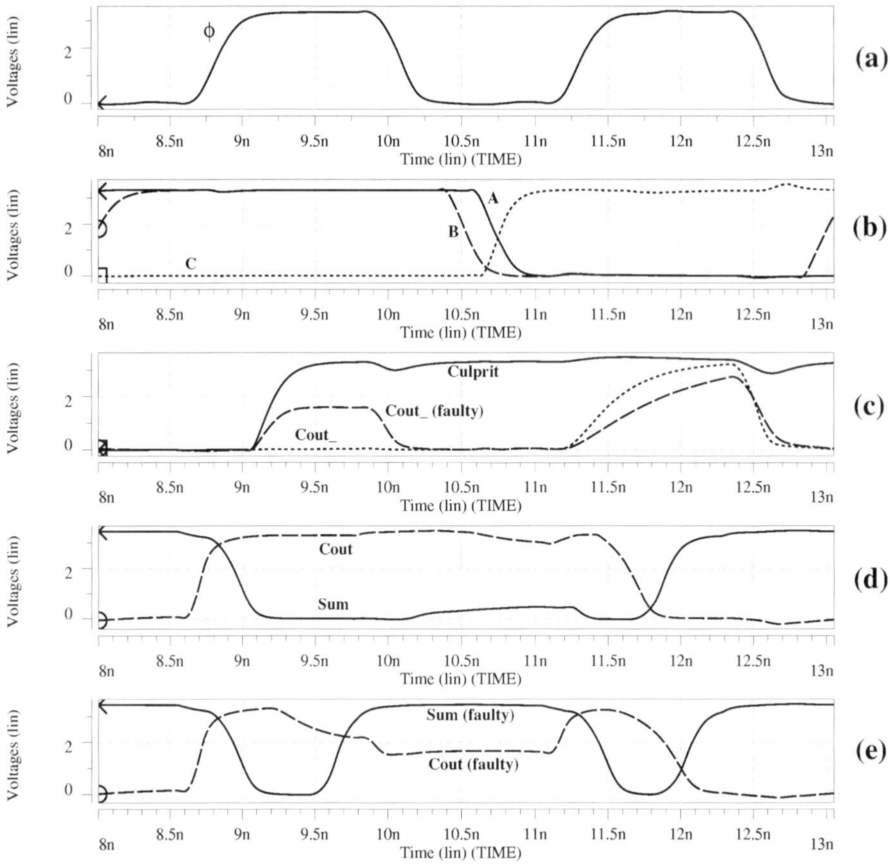

Figure 3.42. HSPICE simulation of the NORA full adder. The effect of transitions of 3 lines coupled to an internal adder line when the latter is in high impedance state is illustrated.

smaller driver resistance there are fewer series transistors. Therefore, the faulty behavior due to coupling in dynamic circuits can be translated as a permanent logic error and not a transient one as in static circuits.

The problems related to crosstalk can be alleviated by the modification of the circuit to include "keeper" transistors, which reduce the impedance in the high impedance state while allowing dynamic behavior. Several dynamic families use this modification [24].

4.2.2 Dynamic RAM cell

Now we will analyze the behavior of a dynamic memory cell with crosstalk interferences. Figure 3.43 shows the structure of the cell. The content of the cell is stored in capacitor C_1, transistors 1 and 2 activate the refresh operation, transistors 3 and 4 select the mode read-write, the sense amplifier is imple-

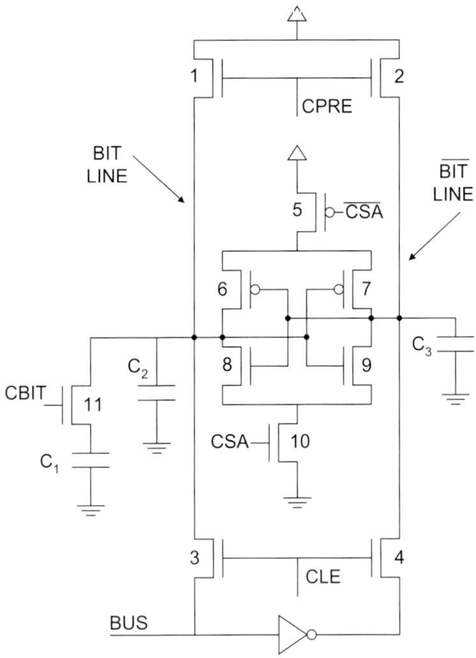

Figure 3.43. Electrical schema of a memory cell structure.

mented with transistors 5, 6, 7, 8, 9 and 10, and the address circuitry drives transistor 11.

The high crosstalk susceptibility of this circuit is due to capacitor C_1 being in a high impedance state while the cell is in storage mode. During this period of time, crosstalk may produce a change in the voltage stored in C_1 due to interconnect activity. If this change is large enough to reach the switching threshold voltage of the technology, then the content of the cell can be permanently changed. Figure 3.44 shows the simulated waveforms in the case of correct behavior, while figure 3.45 shows the situation in which crosstalk causes a change in the voltage of C_1 large enough to produce an erroneous state. In these simulations, a logic 1 is first stored in the cell, and then this value is read as 1 in the correct behavior, and as 0 in the erroneous one. It can be seen in figure 3.45 that the voltage stored in C_1 decreases due to commutation in a near line coupled with node C_1.

Concerning the behavior of this cell, it can be pointed out that another critical state when crosstalk can be dangerous is the time after the pre-charge of the bit line and before the charge is transferred from C_1 to the bit line. In this interval, small changes in the bit line voltage can produce the reading of an erroneous state at the output.

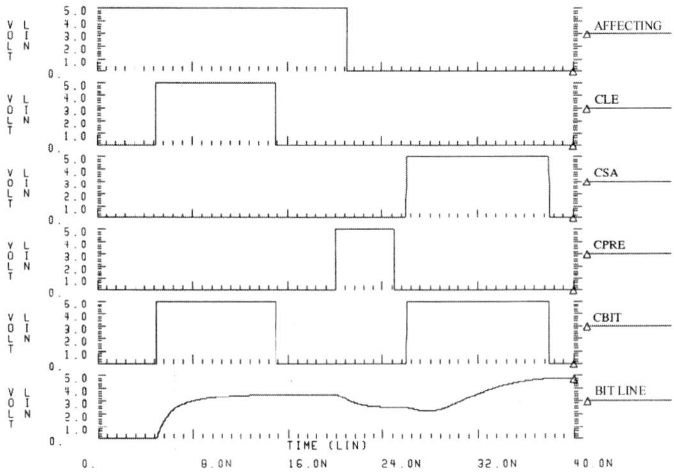

Figure 3.44. HSPICE simulation of the memory cell. Correct behavior.

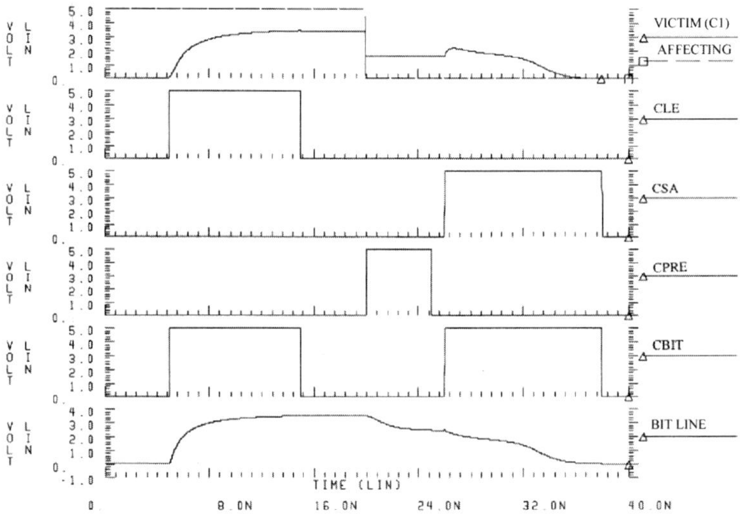

Figure 3.45. HSPICE simulation of the memory cell. Faulty behavior.

References

[1] Roca M., Moll F., and Rubio A. "Electric Design Rules for Avoiding Crosstalk in Microelectronic Circuits". In *4th International Workshop on Power an Timing Modelling for Performance of ICs*, 1994.

[2] Moll F. and Rubio A. "Spurious Signals in Digital CMOS VLSI Circuits: A propagation Analysis". *IEEE Trans. on Circuits and Systems–II*, 39(10), 1992.

[3] Shoji M. *CMOS digital circuit technology*. Prentice-Hall International, 1988.

[4] Sah T. "Characteristics of the Metal-Oxide-Semiconductor Transistor". *IEEE Trans. on Electron Devices*, 11(7), July 1964.

[5] Moll F. *Parasitic effects due to interconnections in microelectronic design*. PhD thesis, Universitat Politecnica de Catalunya, March 1995.

[6] Cartrysse J. "Measured Distortion of the Output-waveform of an Integrated Opamp Due to Substrate Noise". *IEEE Trans. on Electromagnetic Compatibility*, 37(2), May 1995.

[7] Cartrysse J., Sinnaeve A., and Vandecastele G. "Measured Crosstalk on Chips Using Specially Designed Components". *IEEE Trans. on Electromagnetic Compatibility*, 37(2), May 1995.

[8] Soumyanath K., Borkar S., Zhou C., and Bloechel B. "Accurate On-Chip Interconnect Evaluation: A Time-Domain Technique". *IEEE Journal of Solid State Circuits*, 34(5), May 1999.

[9] Caignet F., Delmas S., and Sicard E. "On the Measurement of Crosstalk in Integrated Circuits". *IEEE Trans. on VLSI Systems*, 8(5), October 2000.

[10] Caignet F., Delmas-Bendhia S., and Sicard E. "The challenge of signal integrity in deep-submicrometer CMOS technology". *Proceedings of the IEEE*, 89(4), April 2001.

[11] Delmas S., Caignet F., Sicard E., and Roca M. "On-chip sampling in CMOS Integrated Circuits". *IEEE Trans. on Electromagnetic Compatibility*, 41(4), November 1999.

[12] T. Sato, Sylvester D., Cao Y., and Hu C. "Accurate in situ measurement of peak noise and delay change induced by interconnect coupling". *IEEE Journal of Solid State Circuits*, 36(10), October 2001.

[13] Fourniols J., Roca M., Caignet F., and Sicard E. "Characterization of Crosstalk Noise in Submicron CMOS Integrated Circuits: An Experimental View". *IEEE Trans. on Electromagnetic Compatibility*, 40(3), August 1998.

[14] Werner C., Gottsche R., Werner A., and Ramacher U. "Crosstalk Noise in Future Digital CMOS Circuits". In *DATE 2001*, 2001.

[15] Chandrakasan A. P. and Brodersen R. W. *Low Power Digital CMOS Design*. Kluwer Academic Publishers, 1995.

[16] Sotiriadis P.P. and Chandrasakan A. "Bus Energy Minimization by Transition Pattern Coding (TPC) in Deep Submicron Technologies". In *International Conference on Computer Aided Design*, November 2000.

[17] Kim K.-W., Jung S.-O., Narayanan U., Liu C. L., and Kang S.-M. "Noise-Aware Power Optimization for On-Chip Interconnect". In *Proc. ISLPED '00*, 2000.

[18] Moll F., Roca M., and Isern E. "Analysis of dissipation energy of switching digital CMOS gates with coupled outputs". *Microelectronics Journal*, 34(9):833–842, September 2003.

[19] Avant! Corp. *Star-Hspice 2001.2*, June 2001.

[20] Kim K-W., Baek K.-H., Shanbhag N., Liu C.L., and Kang S.-M. "Coupling-driven signal encoding scheme for low-power interface design". In *International Conference on Computer Aided Design (ICCAD)*, 2000.

[21] Machiarulo L., Macii E., and Poncino M. "Low energy encoding for deep submicron address buses". In *International Symposium on Low Power Electronic Design (ISLPED)*, 2001.

[22] Sotiriadis P.P., Wang A., and Chandrakasan A. "Transition pattern coding: an approach to reduce energy in interconnect". In *European Solid State Circuits Conference*, 2000.

[23] Goncalves N. and de Man H. J. "NORA: A racefree dynamic CMOS technique for pipelined logic structures". *IEEE Journal of Solid State Circuits*, 22(3), June 1983.

[24] Bernstein K., Carrig K.M., Durham C.M., Hansen P.R., Hogenmiller D., Nowak E.J., and Rohrer N.J. *High speed CMOS design styles*. Kluwer Academic Publishers, 1999.

Chapter 4

PACKAGING INTERCONNECTS

As explained in Chapter 1, the development of microelectronic technology and the increase in complexity of integrated circuits have an important impact on package requirements, especially in terms of number of I/O and power dissipation requirements.

This chapter is centered on the electrical aspect of packages and, in particular, on how packages contribute to noise generation. Firstly, a brief introduction to packaging interconnection structure is presented and then several methods are introduced for deriving an electrical model for the package in order to evaluate the overall performance of the chip-package system. Finally, some electrical simulations are presented to illustrate the main effects that packaging interconnects introduce into electrical signals.

1. Packaging structure

In order to make an electrical characterization of packages, a physical and geometrical description is needed. In this section, common package structures are described. From the point of view of electrical connections, several technologies and package families exist, especially concerning external connections (to the printed circuit board, PCB), and connections to the chip.

It is not the purpose of this section to give comprehensive information on manufacturing processes, but to identify common interconnect structures that need to be described in an electrical modeling tool, for example, for obtaining electrical parameters. However, some notions of manufacturing processes are needed in order to understand the different parts of the package, and they are briefly given here.

1.1 Body fabrication technology

There are two families of packages as far as materials are concerned: plastic and ceramic. These determine not only the package's electrical and thermal characteristics, but also the interconnection structure of the package.

1.1.1 Plastic packages

Plastic packages are used for lower performance circuits in which cost is the main driving factor, because they are easy to manufacture and, therefore, cheaper than their ceramic counterparts.

The manufacturing process consists of soldering the chip to a metal frame (leadframe), which will form the package interconnection, and then molding the whole structure in a polymer, usually epoxy [1]. The metal of the leadframe is usually aluminum, or copper covered in gold to avoid chip contamination [2]. This process implies that only one interconnection layer exists and there is no reference plane, so the electrical characteristics of the interconnections are not very good. Only the die attach pad can be considered as a reference plane for a limited area close to the chip. Otherwise, the PCB must provide the reference plane. One exception is the case of plastic BGA packages (PBGA) in which the leadframe is replaced by an epoxy glass substrate to which the chip is attached and connected. After that, it is molded in plastic. In this latter case, the substrate may include several routing layers and reference planes, although usually only bottom and top layers are used for reasons of cost.

The advantages of plastic packages are a low dielectric constant (around 5), which gives rise to a low capacitance of interconnections and a higher velocity of propagation of signals. Another advantage is that their thermal expansion coefficient is very similar to the one used in printed circuit boards (PCB), and mechanical stress is therefore avoided when the temperature of the circuit rises.

On the other hand, the disadvantages are low thermal conductivity, which makes them unsuitable for applications where a large amount of power is dissipated, and the already mentioned non-existence of a reference plane in the package.

1.1.2 Ceramic packages

The alternative to plastic packages are ceramic ones, whether they are all ceramic or have only a ceramic substrate with upper levels of other dielectric materials [3].

The most common technology is the one known as doctor blading [1], [4], in which a ceramic solution is deposited on a polyester film and dried. In this way, a thin and flexible layer of ceramic is obtained, which can be cut into pieces. On each of these layers, a thick-film metallization process is applied and then several ceramic layers are stacked, with fine alignment. They are then pressed and sintered in a thermal cycle reaching 1500° C. This process leads to a great volume reduction, of about 40%, and therefore, it is not possible to achieve fine dimensions for thick-film metallizations.

The materials used for conductors are refractory metals (molybdenum or tungsten) for the ceramic substrate that have the disadvantage of high resistivity (5.3$\mu\Omega$·cm). It is also possible to deposit several thin-film metallization layers on top of the ceramic substrate [5], [6]. In this case, copper (1.7$\mu\Omega$·cm) or aluminum (2.8$\mu\Omega$·cm) are used.

One of the advantages of ceramic substrates is the possibility of using a large number of metallization layers, (up to 150 [5]) and thereby achieving a very high interconnection density, which makes them very suitable for MCM applications. In addition, their thermal conductivity is very high and they can dissipate large amounts of power.

Their disadvantages are a high dielectric constant (around 10) and a thermal expansion coefficient very different from the material (epoxy) used in PCBs. This causes a big problem in circuits with high power dissipation and surface mounting technology, because the circuits can become detached from the PCB due to the mechanical stress caused by thermal expansion.

In order to reduce the problem of the high dielectric constant, the thin-film metallization on top of the substrate may use low-k dielectrics, such as polyimide or silicon dioxide. This has the additional advantage of increasing interconnection density because the thin-film process allows smaller interconnection dimensions.

1.2 External connections

As regards external connections, the traditional classification of packages is based on the method of connecting the package to the PCB, through-hole mounting (THM) and surface mounting (SMT). From the point of view of geometric structure, a different classification can be made between packages with connections on their edges, and packages with connections within their area. Both approaches (edge and area) have versions of THM and SMT packages. Figure 4.1 shows some of the most commonly used package types.

1.2.1 Periphery connection packages

These packages are simpler in construction than the area connection packages and therefore cost less. They show only one layer of interconnection routing, although the body material can be plastic or ceramic.

In this type of package, we can make a further classification depending on the number of sides which have electrical connections: one, two or four sides. As the requirements for connection density at PCB level increase, packages with connections on all four sides are more common and the lead or pin pitch decreases.

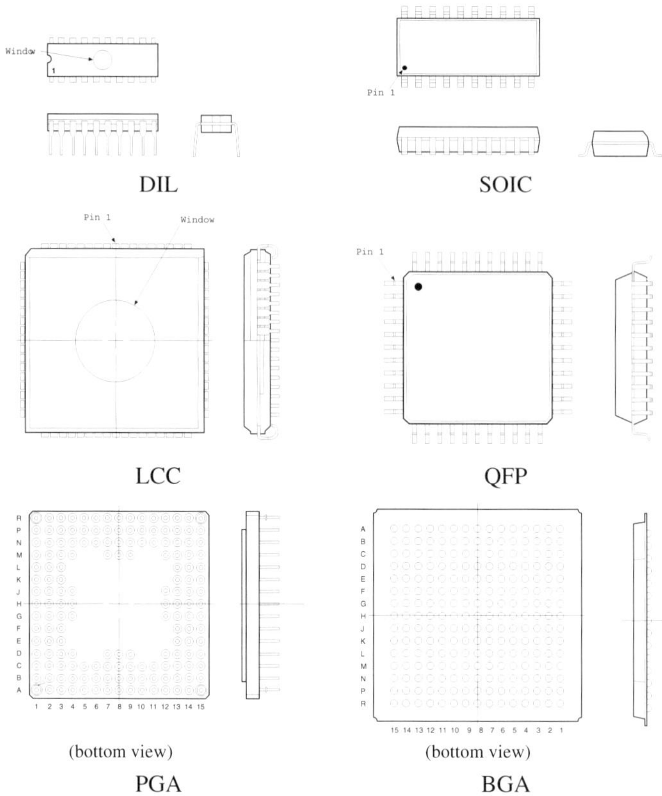

Figure 4.1. Types of packages in terms of their external connection.

Without trying to be comprehensive, the types of packages in this group are known by the following acronyms:

SIP/SIMM Single Inline Package/Single Inline Memory Modules are packages designed to achieve a high component density at the PCB level, by placing the components vertically. They therefore have leads on only one side of the package.

DIL/DIP Dual InLine/Packages are rectangular packages with the leads on two opposite sides, to be soldered into the holes of the PCB (THM technology). The separation between leads in these packages is usually 100 mils ($1\,\text{mil} = 0.001\,\text{inches} = 0.0254\,\text{mm}$).

SOIC Small Outline Integrated Circuits also have leads on only two sides of the package, but they are designed for surface mounting and have a smaller lead pitch than DIL packages. They therefore have a smaller package body. Leads are either gull-wing type or J-lead type. The J-lead pins occupy a smaller area on the PCB, but they generally need a larger pitch than gull-wing leads.

LCC/QFP Leaded Chip Carrier/Quad Flat Packages have connections on all four sides of the package. LCCs have J-lead pins and QFPs have gull-wing pins. QFPs have the greatest connection density among periphery packages, achieving almost 500 leads for a 40 × 40 mm at a 0.3 mm pitch.

1.2.2 Area connection packages

In order to further increase the number of I/Os and connection density at the PCB level, grid array packages were developed. These packages have connections in a whole area of the package substrate instead of only at the periphery. Routing is much more complex than periphery packages and for high performance packages with many I/O terminals, several routing layers and power reference planes are integrated in the package substrate. Both ceramic and plastic packages are available.

There are two families, depending on the PCB mounting technology:

PGA Pin Grid Array packages have leads for THM technology. These are the highest lead count packages available, of the order of 2000.

BGA Ball Grid Array packages are a development of PGA packages, with metallized and plated pads for surface mounting instead of pins. They allow a smaller lead pitch than PGAs.

1.3 Chip connection

The third characteristic concerning package structure—and one which has an important influence on its electrical performance—is the method used to connect the chip to the package.

1.3.1 Wire bonding

Wire bonding (figure 4.2) is the traditional method for chip connection (it was developed at Bell Labs in the late 1950s [7]). It is still the most widely used technique today, especially for low-cost applications. It consists of connecting the chip and package pads with a thin gold or aluminum wire (with a diameter of the order of 20 μm) by a thermosonic or thermocompression bonding process [8]. All the connections must be made sequentially, but, with automated equipment, the time taken for each connection can be as low as 100 ms.

The two main drawbacks of this technique are, firstly, that all connection pads on the chip must be located at on the edges, so the high I/O count increases the chip area and therefore its cost. Secondly, as the wire takes a loop form, it has a relatively high inductance, which degrades signal integrity.

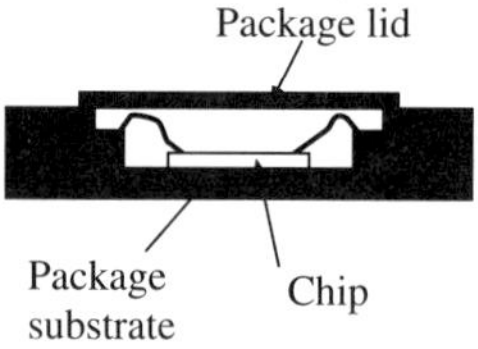

Figure 4.2. Wire bonding chip connection.

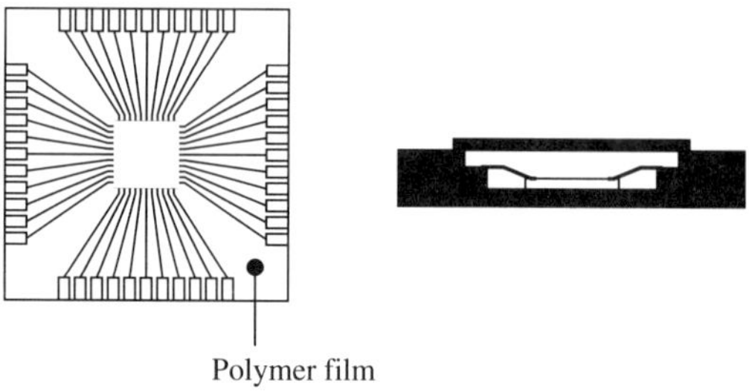

Figure 4.3. TAB type chip connection.

1.3.2 TAB

Tape Automated Bonding is a development of wire bonding. It consists of printing the interconnections on a polymer film (figure 4.3) where the chip and package pad positions are determined. The film is then placed over the chip and solder joints are made using previous deposits of solder bumps on the chip and package pads. After that, the chip and film are attached to the package and connections made to the package bond pads. The details of the process can be found, for example, in [8].

This method provides shorter connections to the package, with a consequent reduction in inductance. The method also allows all the bonds of the chip to be made simultaneously, as well as all the bonds in the package.

1.3.3 Flip Chip

Flip chip (shown in figure 4.4), also known as C4 (Controlled Collapse Chip Connection, developed by IBM in the 1960s) is the equivalent of area connection at chip level. It consists of putting small solder bumps directly on the chip pads. The chip is then placed face to face with the package surface (hence the name flip-chip) and bonds are formed by thermocompression or ultrasonic processes.

This method provides very low inductance and, therefore, maximum electrical performance. In addition, the chip pads need not be at the periphery, but can be scattered all over the chip surface, giving maximum interconnect density.

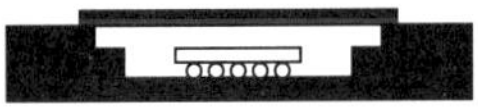

Figure 4.4. Flip chip connection.

The main disadvantage of this approach is reliability and testing, as well as cost (related to yield).

2. Lumped electrical parameter modeling of packages

The modeling problems for packages are the same as those found in integrated circuits. The only difference is in magnitude of the different electrical parameters of lines, drivers and loads. At the package level, interconnection modeling is usually a lumped RLC model [9], [10], especially for single-chip carriers. However, the high switching speed of high-performance circuits may need a transmission line modeling [11], [6]. The power supply lines are especially important because of simultaneous switching noise. For this problem, the most common model is a lumped inductance one [6], [12], [13], [14].

2.1 Calculation of circuit parameters

The tools used to extract circuit parameters from a 3D geometric description of the package conductor can be divided into two families. The first obtains resistance and inductance from different input and output ports, corresponding to the package I/O ports. This method is called Partial Element Equivalent Circuit (PEEC) [15] and it allows the skin effect to be taken into account, so the parameters are frequency-dependent. To obtain capacitance, Finite Element (FEM) or Boundary Element (BEM) methods are used.

As capacitance is calculated separately from resistance and inductance, there is some degree of inaccuracy, unless the conductors are finely discretized. In this case, a set of partial resistances, inductances and capacitances is obtained for each lead-wire conductor. At the limit of infinitessimal discretization, it can be shown that this method is equivalent to a full-wave solution to the problem.

The most important phenomenon related to power supply distribution is the so-called switching noise [13]. It consists of a power supply fluctuation at the chip connection due to the complex impedance of the physical power supply distribution. When the chip node switches, a current demand is created and the impedance causes a voltage variation, which affects all the elements connected to the internal power node, even when only part of the chip is actually switching.

This phenomenon is related mostly to the inductance of the conductors forming the power supply distribution network. When a current derivative appears, the power supply voltage varies. The resistance also causes a voltage variation because of constant current consumption, but this parameter is more easily controlled because the network is designed to have a low resistance. On the other hand, parasitic capacitance is also important because it creates a resonant circuit, together with the inductance which cause voltage oscillations around the DC power supply voltage.

2.2 Concept of Partial Inductance

Inductance as a physical concept has a meaning only in relation to a loop of current (what is referred to as loop inductance). Its value is defined as the relation of magnetic flux through the area defined by the loop and the magnitude of the current causing the magnetic flux, which can be expressed as [16]:

$$L_{ij} = \frac{1}{a_i a_j}\frac{\mu}{4\pi}\oint_i \int_{a_i} \oint_j \int_{a_j} \frac{d\mathbf{l}_i \cdot d\mathbf{l}_j}{r_{ij}} da_i da_j \tag{4.1}$$

being i and j two closed loops, $\mathbf{l}_i$, $\mathbf{l}_j$ their tangent vectors, and a_i, a_j the cross-sectional area of the conductors forming the loop.

Ruehli [16] introduced the concept of partial inductance by considering a discretization of the loop in rectilinear segments, so that inductance is obtained by a summation instead of closed line integrals:

$$L_{ij} = \sum_{k=1}^{K}\sum_{m=1}^{M} \frac{1}{a_k a_m}\frac{\mu}{4\pi}\int_{b_k}^{c_k}\int_{a_k}\int_{b_m}^{c_m}\int_{a_m} \frac{d\mathbf{l}_k \cdot d\mathbf{l}_m}{r_{km}} da_k da_m \tag{4.2}$$

where loop i is divided in K segments while loop j is divided in M segments, each segment with constant cross-section.

This last expression enables a *partial* inductance to be defined for each segment, so that the loop inductance is the summation of the partial inductances of each segment forming the loop:

$$L_{p_{km}} = \frac{1}{a_k a_m}\frac{\mu}{4\pi}\int_{b_k}^{c_k}\int_{a_k}\int_{b_m}^{c_m}\int_{a_m} \frac{d\mathbf{l}_k \cdot d\mathbf{l}_m}{r_{km}} da_k da_m \tag{4.3}$$

$$L_{ij} = \sum_{k=1}^{K}\sum_{m=1}^{M} L_{p_{km}} \tag{4.4}$$

In this way, an inductance value is assigned to an individual piece of conductor, but a physical inductance is not present until several conductors are put together to form a closed loop. Each piece of conductor has a self partial inductance and mutual partial inductances with other conductors depending on their relative distance and orientation. These partial inductances can be directly used in a circuit simulator to obtain a realistic electrical model based on loop geometry.

However, it is important to note that to obtain a realistic response, the physical description must be a quasi-closed loop, because otherwise an important

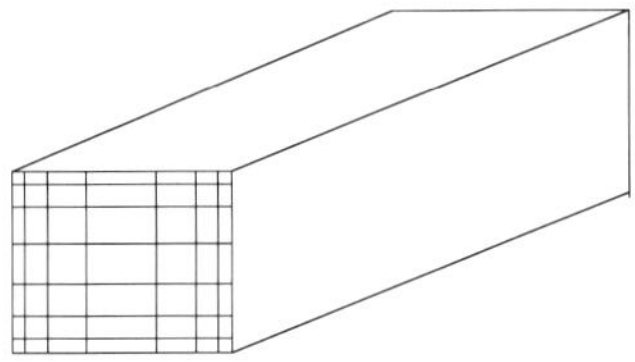

Figure 4.5. Cross-section discretization of a circuit segment to calculate the skin effect.

part of the contribution to the loop inductance (the only physically meaningful one) is missing. For example, a package model with partial inductances may not be useful because the whole loop corresponding to the external power distribution is not described and therefore not taken into account in the model. Therefore, this kind of model must be created with care and its applicability must be carefully analyzed before its use.

2.3 Skin effect calculation

Another feature of the PEEC method is that it allows resistance and inductance values to be considered in function of frequency. At high frequencies, the current density distribution is concentrated at the conductor surface, due to the finite conductor conductivity. This phenomenon is known as skin effect [17].

This effect can be calculated in the PEEC scheme by performing a cross-section discretization, as shown in figure 4.5. This way, each conductor segment is composed of a number of filaments. The current in each filament is calculated by assuming a uniform current density in each filament and taking into account the interaction between different filaments of the same segment. After obtaining the current and voltage for each segment, it is possible to calculate its impedance, that is, resistance and inductance. This method is implemented in a well-known public domain tool, Fasthenry, developed at MIT [18].

2.4 Resistance and inductance calculation

The PEEC method assigns a value for partial resistance and inductance for each segment in the circuit, as well as a capacitance between individual segments. The complete circuit model is equivalent to the electromagnetic model, if retardation effects (because of finite velocity of propagation of EM waves) can be neglected [15]. Retardation effects can also be included in the formulation. In this case the results are equivalent to a full-wave solution [19].

However, the resulting description can be too complex for a typical geometry and, therefore, the corresponding circuit simulation may be too slow to be practical. A simplification of the model consists of finding a relation between physical I/O node currents and voltages at a given frequency. The real part of such a relation will be the resistance and the imaginary part divided by frequency will be the inductance. With this simplification, only one resistance and self-inductance, plus mutual inductances, are obtained for each I/O port (figure 4.6).

The calculations involved for obtaining resistance and inductance are briefly sketched here for illustration. A more detailed explanation can be found, for example, in [20].

In terms of partial resistance and inductance elements, one can write for b partial elements (and consequently the same number of circuit branches):

$$[\mathbf{V_b}] = ([R_p] + j\omega[L_p])\,[\mathbf{I_b}] = [Z][\mathbf{I_b}] \tag{4.5}$$

Where $[\mathbf{I_b}]$ and $[\mathbf{V_b}] \in \mathcal{R}^b$ are currents and voltage differences respectively of the different conductor segments (branches), and $[R_p]$, $[L_p] \in \mathcal{R}^{b\times b}$ their partial resistance and inductance matrices.

From network analysis formulation, currents and voltage at n I/O terminals can be related by means of an *incidence matrix*, $[A] \in \mathcal{R}^{n\times b}$:

$$(\mathbf{I_s}) = (A)(\mathbf{I_b}) \tag{4.6}$$

$$(A)^t(\phi_\mathbf{n}) = (\mathbf{V_b}) \tag{4.7}$$

where $[\mathbf{I_s}]$ is the vector of external (I/O) current sources, and $[\phi_n]$ is the vector of all node voltages. Using equation 4.5 it is possible to find the relation between external currents and induced node voltages:

$$[A]^t[Z]^{-1}[A][\phi_\mathbf{n}] = [\mathbf{I_s}] \tag{4.8}$$

From equation 4.8, by setting consecutively unity current source for each I/O port and solving the matrix equation, it is possible to find the corresponding node voltages and compute the induced voltage difference for each port. The real part of this value will be the port's equivalent resistance and the imaginary part divided by ω, its inductance, self and mutual.

Therefore, the problem depends on the solution of the matrix equation 4.8. In order to speed up the solution, iterative methods can be used. In addition, an alternative mesh analysis, as opposed to the nodal analysis sketched here, gives a more simplified solution [20].

With this method, circuit description complexity is greatly reduced, but, of course, some accuracy is lost because the partial capacitances of the segments are neglected. Capacitance can later be added to the model by separately computing the capacitance of the conductors that form the physical structure.

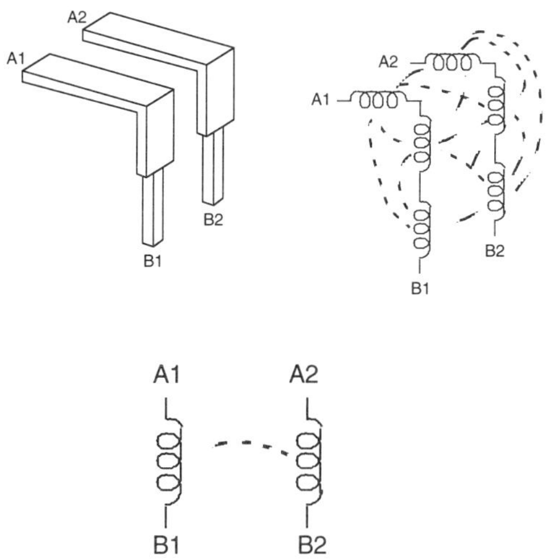

Figure 4.6. Example of complete modeling and simplified modeling of two package leads with partial inductances. Resistances are omitted from the drawing for the purpose of clarity. Dashed lines indicate inductive coupling.

3. Circuit modeling from measurement

Experimental characterizations of packages are needed for two purposes: one is to validate the calculations of parameters obtained by simulation. The second reason is in order to give an evaluation figure for the package, so it can be compared with other packages.

This section briefly reviews several measurement methods that can be found in the literature. For more specialized information, see for example [21] or [22].

In order to derive an equivalent circuit, the measurements performed find the relation between voltage and current using different techniques. To determine capacitance, an open circuit configuration is used and, for the inductance, a current loop must be defined, so the configuration used is a connection to a common node for all the leads. We will see that the measurement set-up has a great influence on the absolute value of the measured inductance. However, the same response can be obtained when introducing the measured values into an electrical simulator, even with different set-ups. On the other hand, for comparing different packages it is very important to define a standard measurement set-up so the measurement results can be fairly compared.

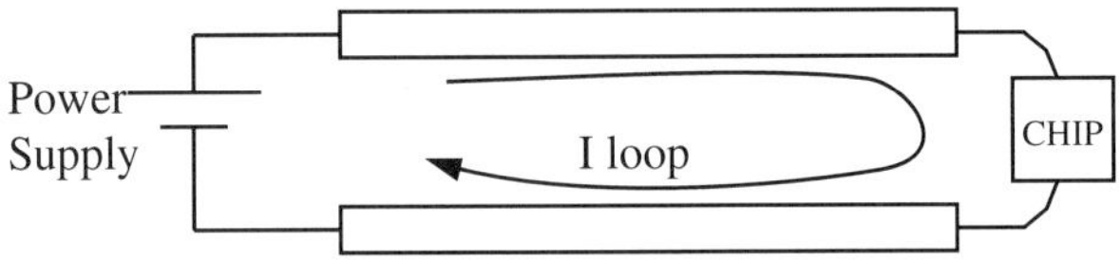

Figure 4.7. A simplified package (only two leads) provides current to a switching circuit from an external power supply.

3.1 Influence of measurement set-up on measured inductance

The partial inductances of a given current loop are not unique, as shown by, for example, Young [23]. They are a useful mathematical artifice for describing the physical element—loop inductance. Given a certain current loop, of all the possible partial inductances, the ones obtained by the PEEC method are equivalent to loop inductances with the return current path located at infinity [16] and they are therefore uniquely defined.

In the case of measurement, the measurement set-up defines a current loop for each of the leads to be measured and, of course, what is obtained is the corresponding loop inductance. Depending on the particular loop or measurement set-up, the inductance obtained will be different.

It can be shown that regardless of the particular set-up and defined loop, and despite the fact that inductances obtained are different, the results of using such inductances are always the same provided there is a common return current path for all the leads in the package.

To show this, let us consider a simplified case: a two-lead package. In order to measure the inductive effect of a circuit connected to the power supply through the two leads, as shown in figure 4.7, two alternative methods can be devised:

1 To define the current loop as the two leads, shorting the inner leads and measuring the relation between external voltage and current (figure 4.7). The current loop is uniquely defined and so is the obtained inductance.

2 To measure each lead separately using a common return path, obtaining a self inductance for each lead and a mutual inductance between them (figure 4.8). The values obtained depend on the measurement set- up (basically, the chosen return path).

The point to be proved is that any approach will give the same result when trying to evaluate the voltage-current relation.

To see that, let us assign to each segment of the circuit (package leads and rest of the current path) its partial inductance as uniquely defined by PEEC method (see figure 4.8). The return current path will have a self inductance

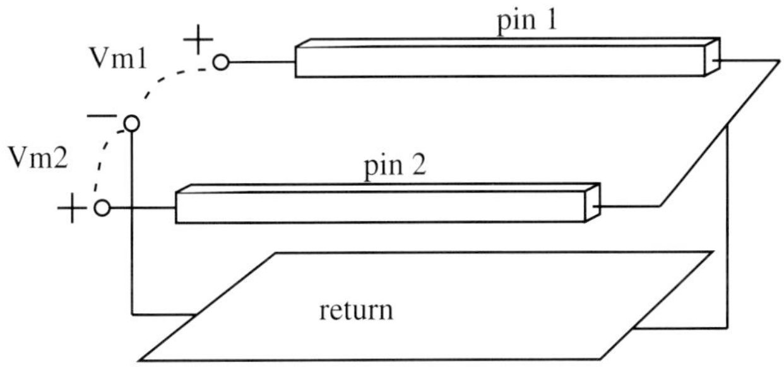

Figure 4.8. Schema of the measurement setup. Each segment can be assigned a partial inductance.

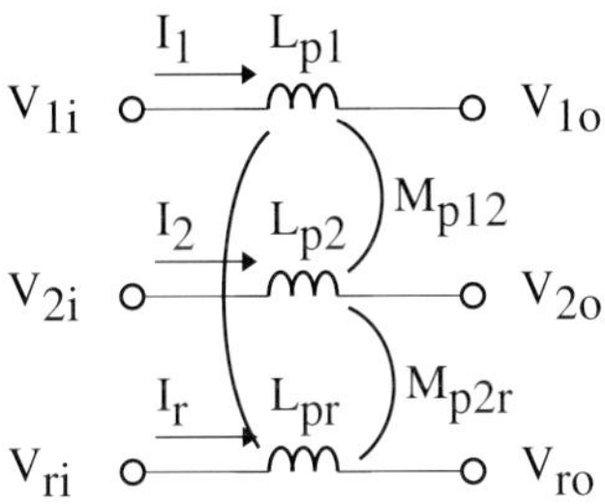

Figure 4.9. Three ports and their partial inductances.

L_{pr} and mutual inductances to the two leads that will depend on the particular path used in the measurement setup. Therefore, the relation between currents and voltages for the three-port system (the two leads plus return lead), as shown in figure 4.9 is:

$$\begin{aligned} V_{1i} - V_{1o} &= j\omega(L_{p1}I_1 + M_{p12}I_2 + M_{p1r}I_r) \\ V_{2i} - V_{2o} &= j\omega(L_{p2}I_2 + M_{p12}I_1 + M_{p2r}I_r) \\ V_{ri} - V_{ro} &= j\omega(L_{pr}I_r + M_{p1r}I_1 + M_{p2r}I_2) \end{aligned}$$

When inductances are measured, the ports labeled o are connected together and voltage relative to V_r is measured, so that, from the previous expressions, $V_{1o} = V_{2o} = V_{ro}$, and $I_r = -I_1 - I_2$ (figure 4.10). The voltages obtained, V_{m1} and V_{m2} can thus be expressed in function of the partial inductances as:

$$\begin{aligned} V_{m1} = V_{1i} - V_{ri} &= j\omega((L_{p1} + L_{pr} - 2M_{p1r})I_1 + \\ &\quad (L_{pr} + M_{p12} - M_{p1r} - M_{p2r})I_2) \\ V_{m2} = V_{2i} - V_{ri} &= j\omega((L_{p2} + L_{pr} - 2M_{p2r})I_2 + \\ &\quad (L_{pr} + M_{p12} - M_{p1r} - M_{p2r})I_1) \end{aligned}$$

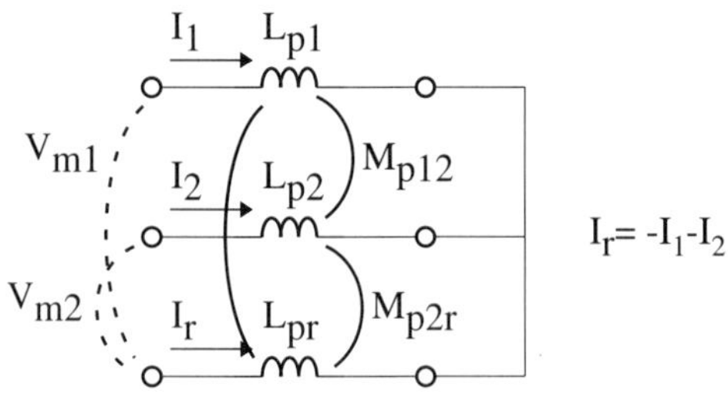

Figure 4.10. Partial inductances in the measurement setup to obtain measured loop inductances.

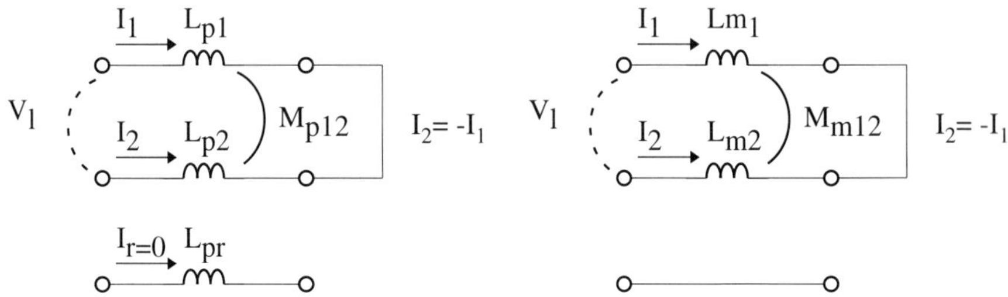

Figure 4.11. Loop inductance calculation through leads 1 and 2 using PEEC inductances and measured inductances.

From the above expressions it is clear that different return current paths (giving different L_{pr}) will give different measured inductances for the same package.

However, if the we analyze the original problem of inductance for the configuration of figure 4.7, we can do it for both approaches, partial inductances and measured loop inductances (figure 4.11). For partial inductances, we have that $I_r = 0$, and $I_2 = -I_1$:

$$V_l = j\omega(L_{p1} + L_{p2} - 2M_{p12})I_1 \tag{4.9}$$

If, instead, the measured values in function of partial inductances are used:

$$\begin{aligned}
V_l &= j\omega(L_{m1} + L_{m2} - 2M_{m12})I_1 = \\
&= j\omega(L_{p1} + L_{pr} - 2M_{p1r} + L_{p2} + L_{pr} - 2M_{p2r} - \\
&\quad 2(L_{pr} + M_{p12} - M_{p1r} - M_{p2r}))I_1 \\
&= j\omega(L_{p1} + L_{p2} - 2M_{p12})I_1
\end{aligned}$$

That is, the result is the same for both cases and is independent of the return current path used in the measurement. As a conclusion, if one wants to compare the measurements from different packages, it is very important that the return path configuration is the same for all the measurements. If, instead, one is interested in using the measured values for package evaluation, for example by using the values in a circuit simulator, then the return path location is not important.

3.2 Time domain measurement methods

There are two types of measurements from which an equivalent circuit can be derived.

3.2.1 Measurements with ramped stimuli

This method was presented by Quint et al. [10]. It derives a lumped capacitance by measuring the voltage induced when a ramp input voltage is applied, with the lead in open circuit configuration. The equivalent circuit is shown in figure 4.12. If the ramp is long enough to let the output voltage achieve a steady value, then the value of this steady value is:

$$V_o|_{st} = R_{sc} C_x \frac{\Delta V_{in}}{\Delta t} \tag{4.10}$$

Knowing the input ramp (derivative), the oscilloscope input impedance, R_{sc} and the output voltage, C_x can be determined. This method allows the capacitance between any pair of package leads to be determined by leaving all the leads in the package in open circuit and then driving one lead and measuring the other. Note that what is actually measured is the total capacitance between leads, including the contribution of the equivalent capacitance of the other leads, as shown in figure 4.13.

In order to obtain the value for self inductance, the lead to be measured is connected to ground, forming a loop as shown in figure 4.14. The oscilloscope impedance will normally be much greater than the lead impedance, so the source current is very approximately equal to the current in the lead. Solving for the output voltage, after a transient behavior of time constant $\tau = (R_g + R_x)/L_x$, the response is also a ramp, following the expression:

$$V_o|_1 = K \frac{R_x}{R_g + R_x} t + K \frac{Rg}{R_g + R_x} \frac{L_x}{R_g + R_x} \approx K \frac{R_x}{R_g} t + K \frac{L_x}{R_g} \tag{4.11}$$

where K is the ramp input voltage (time derivative) and where $R_g \gg R_x$ is assumed. After the ramp, with constant input voltage, the output voltage takes a constant value after a transient:

$$V_o|_2 \approx V_{i\max} \frac{R_x}{R_g} \tag{4.12}$$

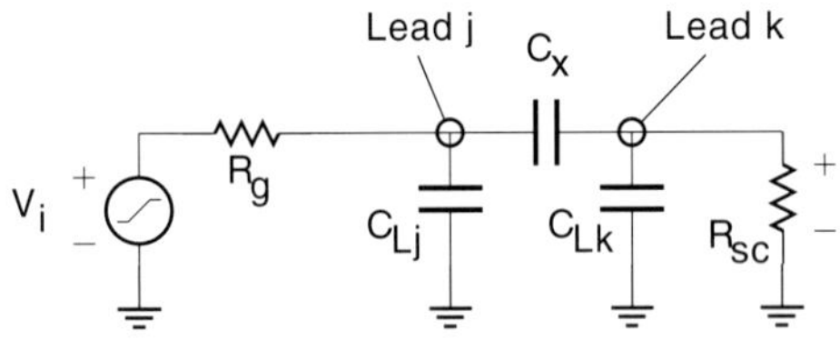

Figure 4.12. Equivalent circuit for measurement of capacitance with a ramped input.

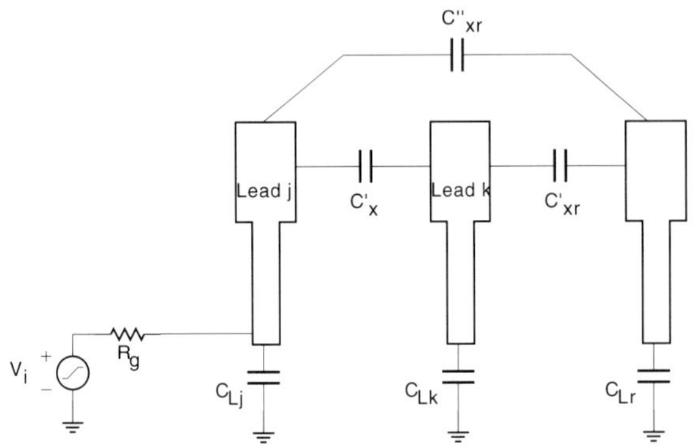

Figure 4.13. Contribution of other leads to measured capacitance.

The value of R_x can then be calculated from this last steady value, knowing the input ramp amplitude V_{imax} and the source resistance, R_g, while L_x can be obtained from the difference between the maximum output voltage and the steady final value, as shown in figure 4.15.

The value of mutual inductance, M_x, can be calculated by measuring the induced voltage in a shorted neighboring lead, as shown in figure 4.16. After a transient, the voltage is constant, while the input increases linearly, achieving a constant value:

$$V_2|_{st} \approx K\frac{M_x}{R_g} \tag{4.13}$$

This technique is simple to implement and does not need very sophisticated equipment. The only requirements are that the oscilloscope must have good resolution and the voltage source must be capable of providing the current demanded by the inductance measurement. The result is a lumped parameter model, which, although it is obtained from low frequency measurements, can be useful up to high frequencies: at least, up to the first resonance frequency.

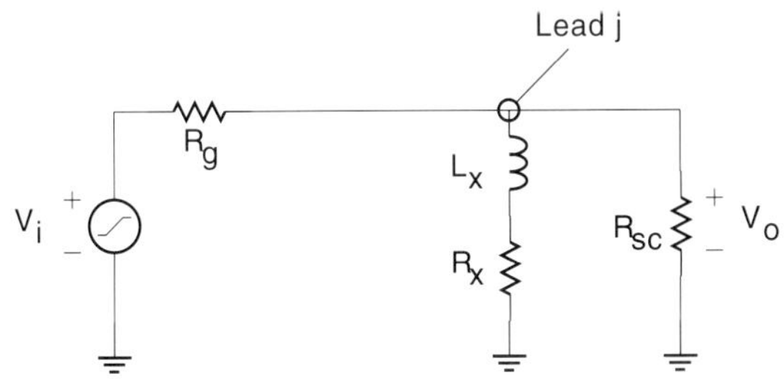

Figure 4.14. Equivalent circuit for measurement of self inductance with a ramped input.

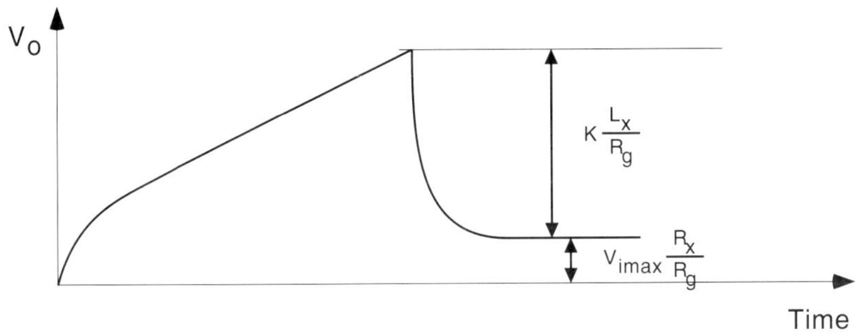

Figure 4.15. Waveform from self inductance measurement with a ramped input.

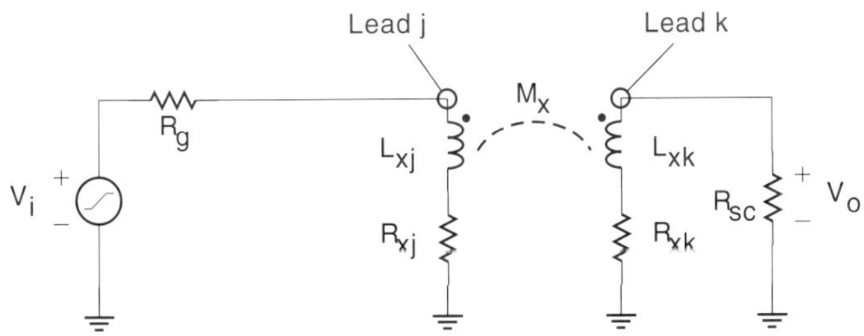

Figure 4.16. Equivalent circuit for measurement of mutual inductance with a ramped input.

3.2.2 Time domain reflectometry

This technique is based on analyzing the voltage waveform obtained from successive reflections of the discontinuities in the interconnect path after a sharp voltage transition is injected. By analyzing the successive voltages, it is possible to derive an *impedance profile*, in function of time measured, $Z_o(t)$. In the case of multiple reflections, the different impedances associated with each path segment can be derived using several algorithms [24], [25] [26]. Figure 4.17 shows the difference between the voltage waveform obtained and the impedance profile.

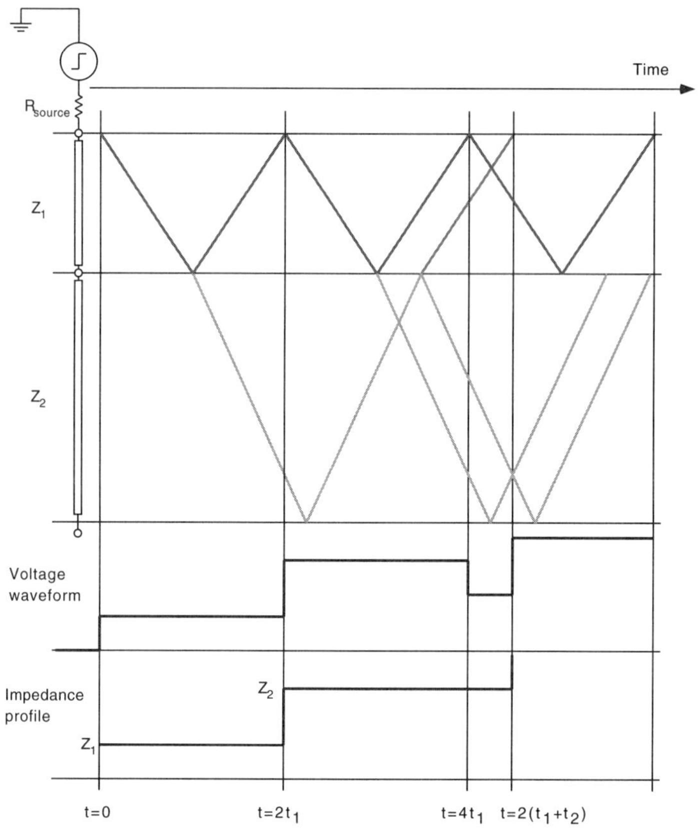

Figure 4.17. Impedance profile obtained from the reflected voltage waveform.

There is a relation between the measured time, t, of the impedance profile and the distance along which the discontinuity is located. This relation is given by the velocity of propagation and line length, or, equivalently, the time of propagation:

$$t = \frac{2l}{v_p} = 2t_p \tag{4.14}$$

This latter expression enables the value of inductance and capacitance to be calculated from the impedance profile obtained. Remembering that for a lossless transmission line, given a length of line l, the total inductance can be obtained as:

$$L_{tot} = Ll = Z_o t_p \tag{4.15}$$

using equation 4.14, for a constant value of impedance, Z_o, the total inductance depends on the time measured corresponding to the time of propagation:

$$L_{tot} = \frac{1}{2} Z_o t \tag{4.16}$$

This last expression can be generalized to the case where Z_o is not constant, by taking small segments of line length and then summing, which leads to an integration between times t_1 and t_2, corresponding to distances from the point of signal injection:

$$L_{tot} = \frac{1}{2} \int_{t_1}^{t_2} Z_o(t) dt \tag{4.17}$$

Similarly, the total capacitance between two distances corresponding to times t_1 and t_2 can be obtained:

$$C_{tot} = \frac{1}{2} \int_{t_1}^{t_2} \frac{1}{Z_o(t)} dt \tag{4.18}$$

In summary, time domain reflectometry is a technique that needs complex post- processing of the measured waveform. In principle, it is not necessary to connect the package leads in a predetermined manner for measuring inductance and capacitance because the technique can identify several segments with their distinct characteristic impedance. However, in order to be able to distinguish the different segments, the rise time of the signal must be much smaller than the signal propagation time through that segment, which is a serious limitation for small packages, as the shortest generated rise times for this type of equipment are on the order of tens of pico-seconds. Otherwise, expressions 4.17 and 4.18 can be used to compute a lumped inductance or capacitance. In this case, the best approach is to use the standard set-up (short-circuit for inductance and open-circuit for capacitance).

3.3 Frequency domain measurement methods

Methods for measuring package parasitics in the frequency domain have the advantage of considering parameters with values dependent on frequency, unlike time domain methods that obtain constant parameter values.

These methods are based on two types of instruments: LCZ meters, for a small frequency range (up to 10 MHz), and Vector Network Analyzers, for measurements up to several tens of GHz. In the second case, an equivalent circuit is derived from the measured S-parameters.

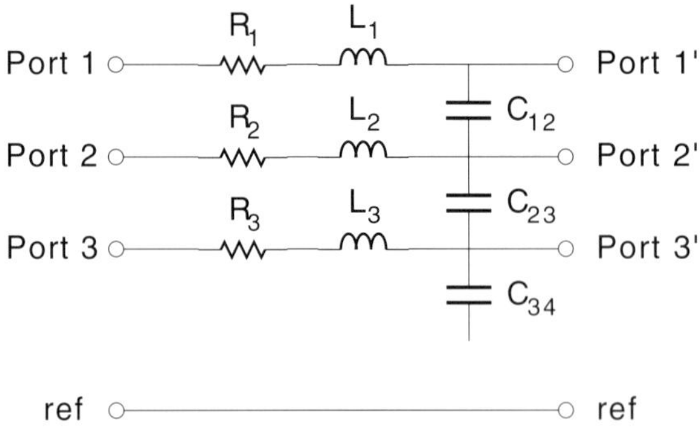

Figure 4.18. Circuit model assumed for comparison with measured S-parameters. Mutual inductances are omitted for clarity purposes.

3.3.1 LCZ meter measurements

The LCZ meter measures complex voltage and current at a certain port and calculates its complex impedance. By configuring the package connection as a short-to-ground or open circuit, its inductance and capacitance can be correspondingly measured. Even though the measurements are performed at a relatively low frequency, the equivalent circuit is valid up to a much higher frequencies, as with the time domain ramped input technique.

3.3.2 Network Analyzer Measurements

This instrument measures the S-parameters of the device under test, in this case an interconnect. Several techniques can be found in the literature based on this principle [27], [28], [29], [30]. Briefly explained, they all assume a certain circuit model with lumped parameters, calculate the circuit's S-parameters for various configurations (at least, open and short) and then they obtain the circuit parameter values comparing the circuit response with the S-parameters measured.

In general, the circuit model is like the one shown in figure 4.18. By connecting all the inner leads to ground, one can assume that all capacitors are shorted and that they do not contribute to measured S-parameters in that configuration. From this set-up, values for inductance and resistance are obtained. If all leads are left open, currents can be assumed to be zero and inductance does not influence the parameters measured. The values for capacitance are then extracted from this second set-up.

On one hand, the measurements are more precise than those obtained by an oscilloscope. However, although it is possible to measure at a high frequency, the assumption of lumped parameter circuit behavior is not valid at such high frequencies. In addition, the complexity of the lumped parameter model is limited by the different configurations that can be used with the package connections.

4. Power distribution modeling from EM simulation

The above sections have explained how to calculate a lumped parameter model for package (or, in general, interconnection) geometry. This model enables electrical simulations to be performed using standard SPICE-like circuit simulators or behavioral simulations, for which the standard approach is IBIS [31]. However, the lumped approximation mode is valid only for a limited range of frequencies.

The electromagnetic simulations compute the electrical and magnetic fields in the structure for a user-defined stimulus and given a certain conductor geometry. The voltage and current at certain spatial positions can later be calculated from the field magnitudes.

Generally speaking, there are two types of solutions for Maxwell's equations: in the frequency domain and in the time domain. Frequency domain tools usually compute S-parameters for the terminals in the structure. This information can be introduced directly into some circuit simulators and the transient response computed by a convolution approach [32], [33]. Time domain simulators obtain transient waveforms for voltage and current at the structure terminals. From these, frequency-dependent magnitudes (typically, impedance) can be obtained by means of the Fourier transform.

While EM simulation of complex packaging structures is a very computer intensive task [34], for very high frequencies, the lumped parameter approximation is no longer valid and a more rigorous model is necessary. Innovative algorithms and computer implementations, together with advances in computer speed capabilities enable the applicability of this approach to be extended for structures of moderate complexity [34].

4.1 Stimulus definition for EM simulation with FDTD approach

The Finite Difference Time Domain (FDTD) approach [35] discretizes the physical structure and solves Maxwell's equations directly in the time domain. It therefore allows the transient response of the electric and magnetic fields to a certain transient stimulus to be obtained directly. In this respect, it is more attractive than frequency domain methods, because it is even possible to link the EM simulation directly with a transient circuit simulation.

One important application field for EM modeling is the calculation of Simultaneous Switching Noise (SSN), which is the voltage fluctuation at a chip's power supply terminals because of a current demand (current derivative) [14].

In order to obtain an approximate response from the FDTD simulation, it is necessary to analyze the phenomenon of the switching event and the way in which the switching current circulates through the package. Two possible switching scenarios must be considered: one is when one or more on-chip nodes are switching (*internal switching*). The other is when one or more package output nodes, or off-chip nodes, switch (*external switching*).

In the case of an internal switching event, there is a current demand from the chip's power supply terminals so that an on-chip node capacitance is charged (or discharged). The total charge associated with this event is:

$$Q = \int i_{sw} dt = V_{DD} \sum_{i=1}^{N} C_{li} \tag{4.19}$$

where i_{sw} is the current supplied for the switching, V_{DD} is the power supply voltage (assuming all the nodes switched are charged to this value), and C_{li} is the capacitance of each node simultaneously switching.

In this situation, switching modeling of a circuit is carried out as shown in figure 4.19. An independent current source (stimulus for the EM simulation) is connected to the internal power terminals of the package, while the external power terminals are connected through a small resistor which accounts for the internal resistance of the power supply. In addition, a capacitor between the current source terminals can be added to emulate on-chip parasitic capacitance. In this way, voltage fluctuations obtained in the internal power terminals result from the simulation of EM waves propagating inside the structure as a consequence of the current source, usually modeled as a gaussian waveform. Once the voltage waveform is obtained, it is possible to compute the Fourier Transform for both waveforms (current stimulus and computed voltage) and then to obtain the impedance of the structure as a function of frequency. This is a useful magnitude for identifying resonance frequencies, which may cause problems if the bandwidth of the signals propagating in the package is near one of these frequencies.

This approach obtains voltage waveforms similar to a realistic case. However, realistic waveforms for voltage switching noise cannot be obtained because the excitation is an independent source, while in reality the current waveforms for devices are, in turn, affected by noise and therefore a certain feedback is produced [14]. Then, instead of this simplified setup, it is possible to couple the FDTD simulation with a standard circuit simulation, including non-linear devices such as the switching elements. This gives realistic results, but it is very expensive in terms of computation resources. Besides, modern technologies demand complex device models which are not usually implemented in the circuit simulators coupled to FDTD tools.

In the case of external switching, the simplified FDTD analysis is not possible because there are several current paths: part of the current goes through

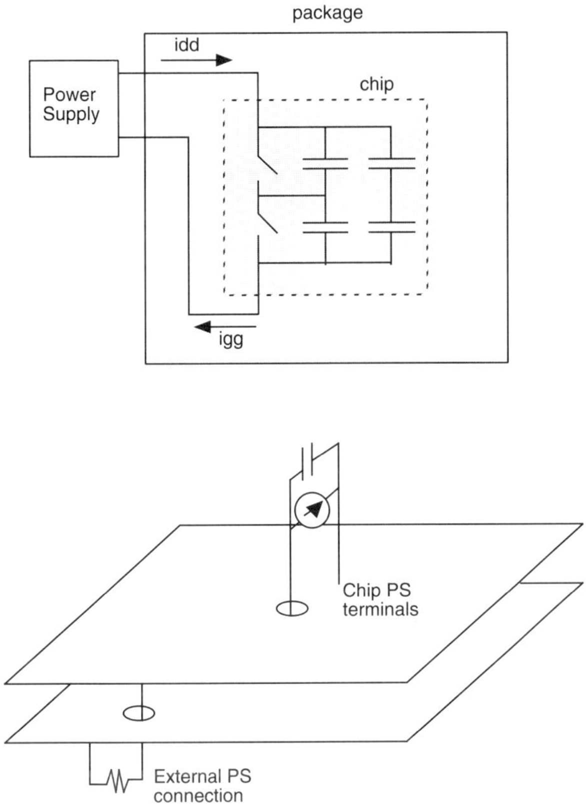

Figure 4.19. Schematic depiction of an internal switching event (top), and the corresponding stimulus definition necessary for EM simulation (bottom).

the signal line to the load, part of it goes from the on-chip positive to the negative power supply terminals. Therefore, it is not possible to consider a single current source stimulus as in the case of internal switching.

In order to model SSN, one can consider coupled FDTD-circuit simulation, or, alternatively, try to "extract" a suitable circuit from the results of the EM simulation such that this circuit model has a validity range beyond that of the lumped parameter calculations presented in section 2. In addition, the model thus obtained can be readily used in conjunction with commercial circuit simulators, which include very advanced and accurate semiconductor device models. This is the approach illustrated in the following sections.

4.2 Circuit model obtained from EM simulations

In order to illustrate this method, let us consider the following problem [36]: to extract a circuit model for the power distribution of a simple package struc-

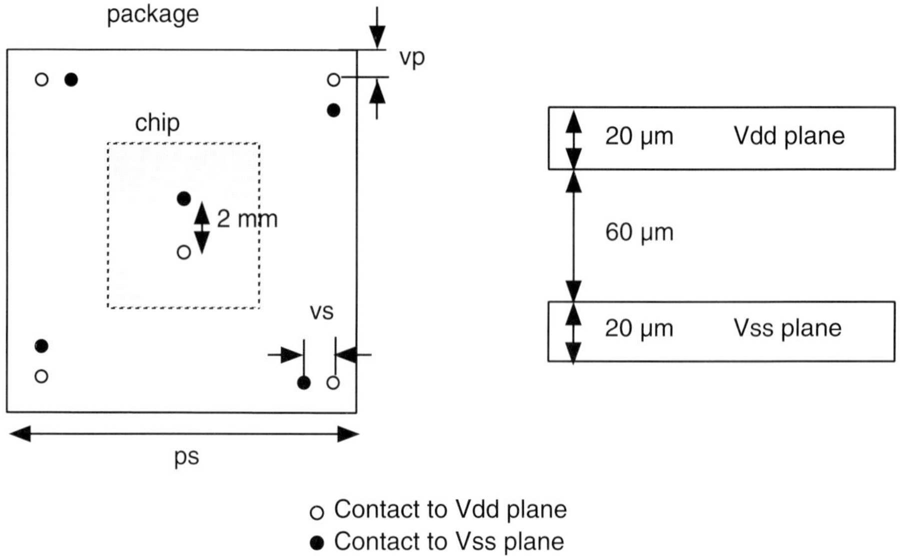

Figure 4.20. Structure and dimensions of the package power distribution used for the EM simulations.

ture consisting of two planes, as depicted in figure 4.20. Two identical structures are simulated, with different dimensions:

pack50 ps: 50 mm, vs: 1 mm, vp: 10 mm.

pack100 ps: 100 mm, vs: 5 mm, vp: 5 mm.

This structure is first simulated with an FDTD tool, SPEED2001 [37]. Then, in order to obtain an equivalent circuit model, HSPICE optimization features are used [38]. Starting from a certain circuit topology defined by the user, the program is able to find the parameter values such that the results match a set of characteristics (for example, peak amplitude, time for zero-crossing, etc.) The characteristics are extracted by observing the EM simulation.

This optimization can be made from time-domain characteristics with a transient circuit simulation, or from frequency-domain characteristics and then performing a frequency domain (.AC) circuit simulation.

The circuit topology used in the circuit simulations for matching the physical structure of figure 4.20 is shown in figure 4.21. It consists of several LC stages.

The results obtained are presented below, with a comparison between transient and frequency domain optimization.

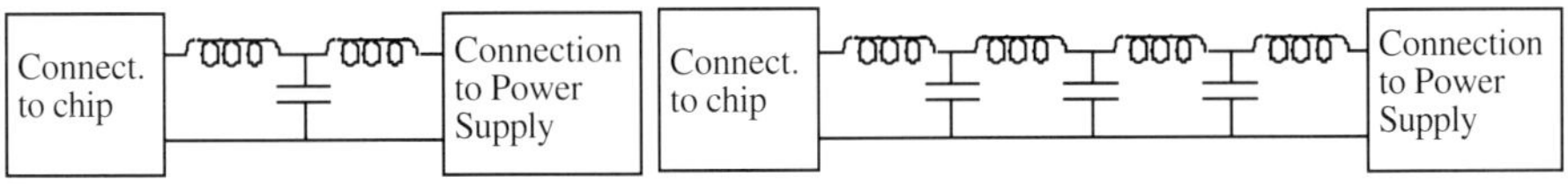

Figure 4.21. Topologies for the equivalent circuit.

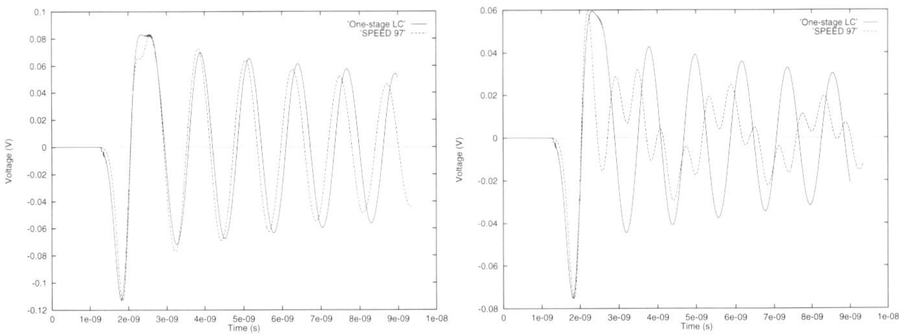

Figure 4.22. Results of both SPEED and HSPICE simulations after optimization of a one LC stage model. Left: *pack50*. Right: *pack100*.

4.2.1 Time domain optimization

The complexity of the circuit topology to be optimized depends on the frequency components of the excitation, as discussed later in more detail. If the switching current has a small bandwidth compared to the resonance frequency of the structure (in other words, it varies slowly with respect to the time of propagation of signals through the structure), then a simple model with a small number of stages is sufficient to model the power distribution accurately. If, on the other hand, the current variation is very fast, then a more complex model will be necessary.

This can be observed in figure 4.22, where two cases are shown: one corresponding to package *pack50* defined above, and the other to package *pack100*. Both are equally excited with a gaussian current waveform of 300 ns width. A simple circuit model of one stage only is considered. The optimization goals are set to the peak voltage of the first, second and third peak, as well as the time of first crossing by zero. For case *pack50* the results for this simple model are fairly good, while for the bigger package, *pack100*, the results are not accurate.

This difference in the success of the optimization process can be explained by looking at figure 4.23. The frequency-domain impedance is represented there for both the SPEED simulation and HSPICE results. For package *pack100*, the first resonance frequency is lower than for *pack50* (because the physical dimensions are greater). As the model considered, with one LC stage, presents

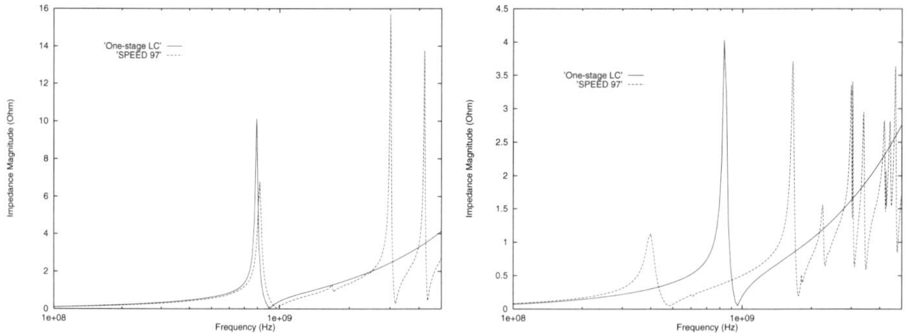

Figure 4.23. Magnitude of the impedance obtained from SPEED and HSPICE after optimization of a one LC stage circuit model. Left: *pack50*. Right: *pack100*.

just one resonance peak, the optimizer obtains a compromise resonance frequency for *pack100*, trying to accomplish the goals of the peaks and zero crossing times, which finally translates in an inaccurate result. It is clearly seen from this figure that either a more complex circuit model is needed or the frequency validity of the one-stage model limited to before the second resonance peak (around 2.5 GHz for *pack50*, and 1 GHz for *pack100*), in which case a stimulus of 300 ns width cannot be applied.

4.2.2 Frequency domain optimization

The approach of performing the optimization in the frequency domain allows better control of the frequency range for which the circuit model is valid. For example, a 3-stage circuit like the one shown in figure 4.21 allows the adjustment of up to 3 resonance frequencies, which in the case of *pack100*, enables the validity of the circuit model to be increased to approximately 3 GHz.

Figure 4.24 shows the optimization result in the frequency domain, taking as goals the values of impedance magnitude peaks, as well as frequencies corresponding to different crossings with zero magnitude. Also shown is the transient simulation with the model obtained from frequency domain optimization. Now, the circuit simulation is much more similar to the EM simulation.

5. Simulations of package effects

As an illustration of the models, some simulations are presented here showing the effect of package contribution to the overall noise inside the integrated circuit [39].

The objective is to see the delay variation due to two causes: crosstalk (simultaneous transitions in neighboring lines, as explained in chapter 3), and package parasitics (mainly, power supply fluctuations due to simultaneous switching noise).

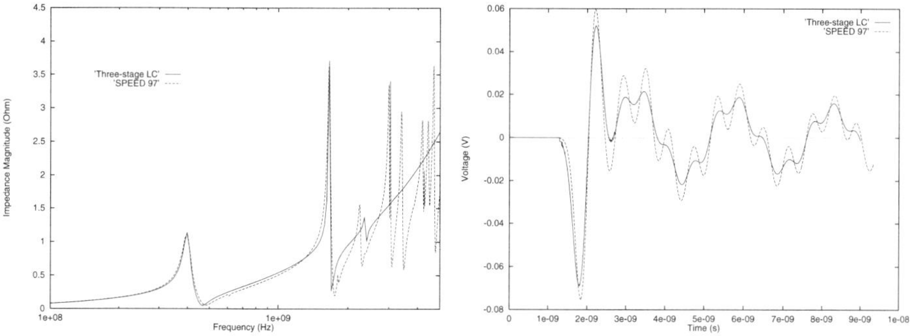

Figure 4.24. Results from SPEED simulation and HSPICE simulation of *pack100* using a 3 LC stage model. Left: frequency response (impedance magnitude). Right: transient response to a 300 ns width gaussian.

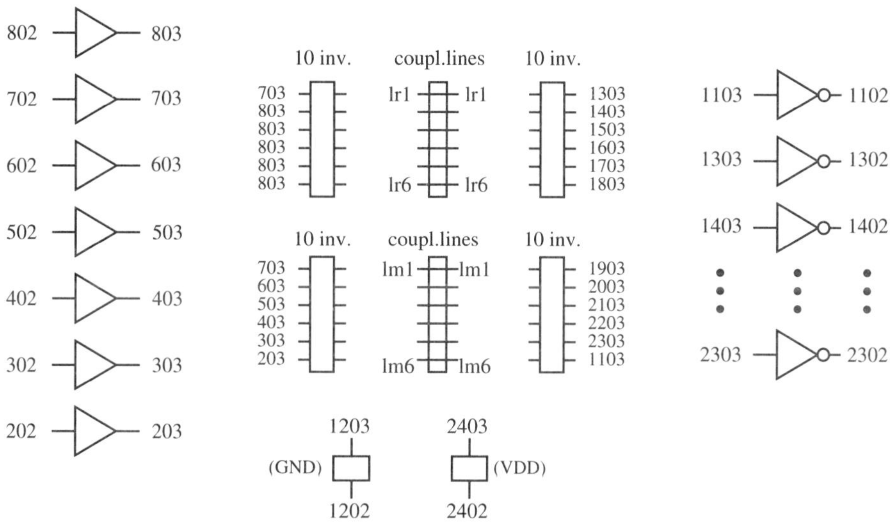

Figure 4.25. Complete circuit simulated, including the two sets of coupled lines, chains of inverters, and I/O pads.

The circuit chosen to see these two effects is shown in figure 4.25. There is a logic path containing 6 coupled chip lines (their vertical structure is shown in figure 4.26). The delay simulated in one of them depends on the activity of the other 5 lines. In order to compare the delays with and without crosstalk, an identical set of 6 lines is included as shown in figure 4.25 so that all the other 5 lines in this second set are quiet. Therefore the delay in this second set serves as a reference delay, without the effect of crosstalk. The delay is "measured" from nodes 703 to 1303 in the reference set, and from nodes 703 to 1903 in the crosstalk-affected set. Before and after the coupled lines, there is a chain of 10

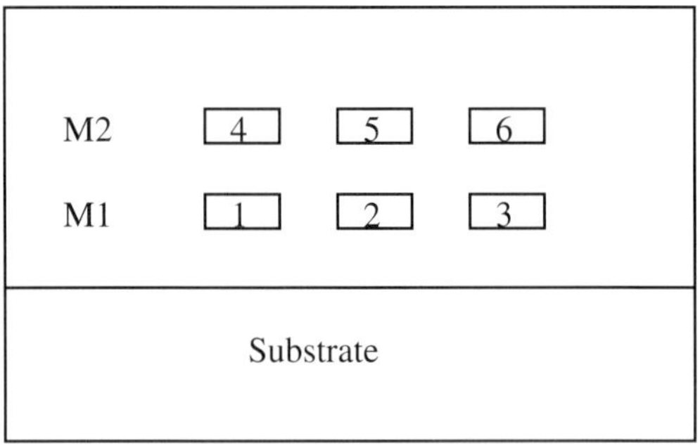

Figure 4.26. Vertical structure of the 6 coupled lines used to evaluate the effect of crosstalk.

inverters for each line so a realistic delay path is simulated. The total number of inverters switching in the worst case is $(6+6) \times (10+10) = 240$. Each of the 12 first inverter chains is driven by a signal provided by input driver pads (7 input pads, corresponding to the 7 different signals needed to simulate the cases of interest), and there is an output driver pad for each of the 12 inverter chains after the coupled lines.

In order to simulate the package parasitics, electrical models of three different packages have been used: DIL24, QFP44 and PLCC44. The circuit delay is, therefore, measured with the package parasitics for each package and without any package parasitics to see how much the delay is affected by package. Figure 4.27 shows the package pin assignments used for each simulation. The signal assignment is similar for all packages to enable an easy comparison between the three of them.

In order to illustrate the effects stated in the introduction, two types of results are presented:

1 How the reference delay (without crosstalk) changes with the rest of the circuit increased activity due to package parasitics (SSN). This is shown in figures 4.28 to 4.31.

2 Changes in delay because of crosstalk, with and without package parasitics (figures 4.32 to 4.35).

There is a different effect depending on the transition polarity of the transitions causing the variation in delay with respect to the transition in the reference line. In all the graphs, the corresponding results are noted as *av* or *rt*, which stand for "advancing" (if both transitions are in the same direction, causing an "advanced delay") or "retarding" (if transitions are in the opposite direction, causing a "retarded delay") [40].

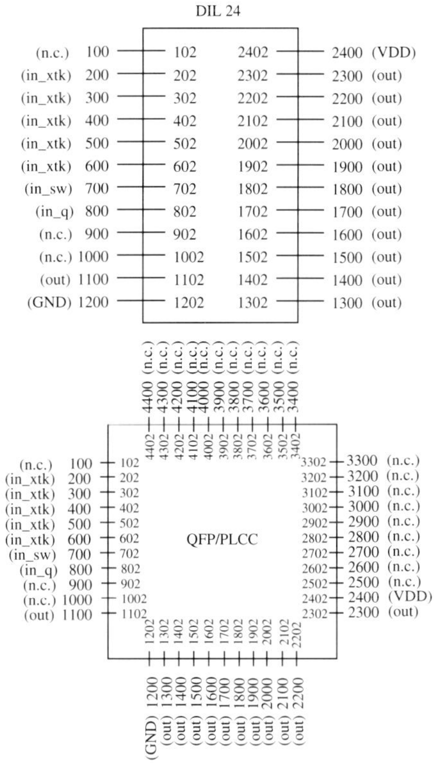

Figure 4.27. Package pin-out used in the simulations. *insw* is the signal causing the transition to be measured, *inxtk* are transitions simultaneous to the previous one, either in the same direction, or the opposite direction. *inq* is a quiet input, connected to zero.

5.1 Effect of SSN

The first set of graphs, in figures 4.28 to 4.31 show the dependence of the delay in the reference set of lines, that is, without crosstalk effect, only due to package parasitics. There are two graphs for each of the lines in the coupled set, taking into account the fact that because of the symmetry in line structure shown in figure 4.26, the results for line 3 and line 6 are equal to those of line 1 and line 4 respectively. Case 1 to 5 means different circuit activity. From one case to the next, circuit activity is increased by 20 inverters plus one input pad and one output pad.

Several observations can be drawn from this set of figures. Firstly, as fully expected, the type of package is a primary influence on SSN. In this respect, the DIL package cause a greater delay variation than the other two packages considered here. The way in which the different victim lines show different path delays depending on the total capacitance they present, can also be ob-

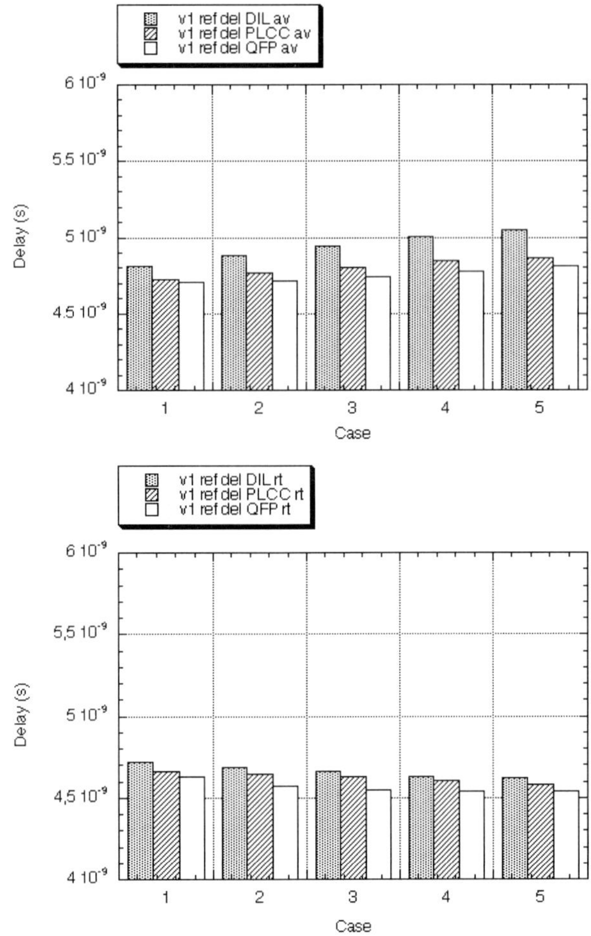

Figure 4.28. Reference delay with line 1 as victim. The lines switching in each case are: 1+2, 1+2+4, 1+2+4+5, 1+2+3+4+5, 1+2+3+4+5+6.

served. In order of increasing delay, they are as follows: line 4, line 5, line 1 and line 2.

Secondly, it can be seen how the polarity of the input transitions cause either an increase or a decrease in a given signal delay inside the chip. This difference in behavior is due to the different paths of switching current when a particular driver switches with a HL or a LH transition. The superimposition of all the current waveforms of each driver gives an overall result, which is very difficult to predict analytically. In general, this overall effect will produce a certain amount of signal jitter when an arbitrary and varying number of signals

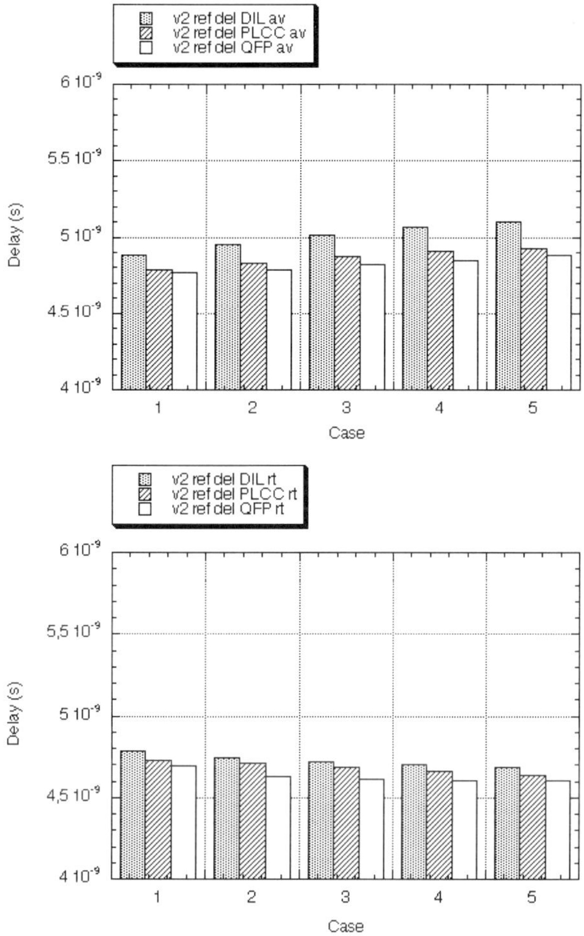

Figure 4.29. Reference delay with line 2 as victim. The lines switching in each case are: 1+2, 1+2+5, 1+2+3+5, 1+2+3+4+5, 1+2+3+4+5+6.

are switching in either direction. The effect of the package must therefore be strictly controlled in order to minimize this SSN-induced jitter.

5.2 Effect of crosstalk

A second set of graphs is presented in figures 4.32 to 4.35. They show the effect of line coupling on path delay, for each of the six coupled lines considered as the victim. In all the graphs, Case 1 corresponds to a single line delay, without crosstalk or SSN effects.

The results shown in these graphs require some discussion for a proper interpretation. The data corresponding to the circuit structure without package

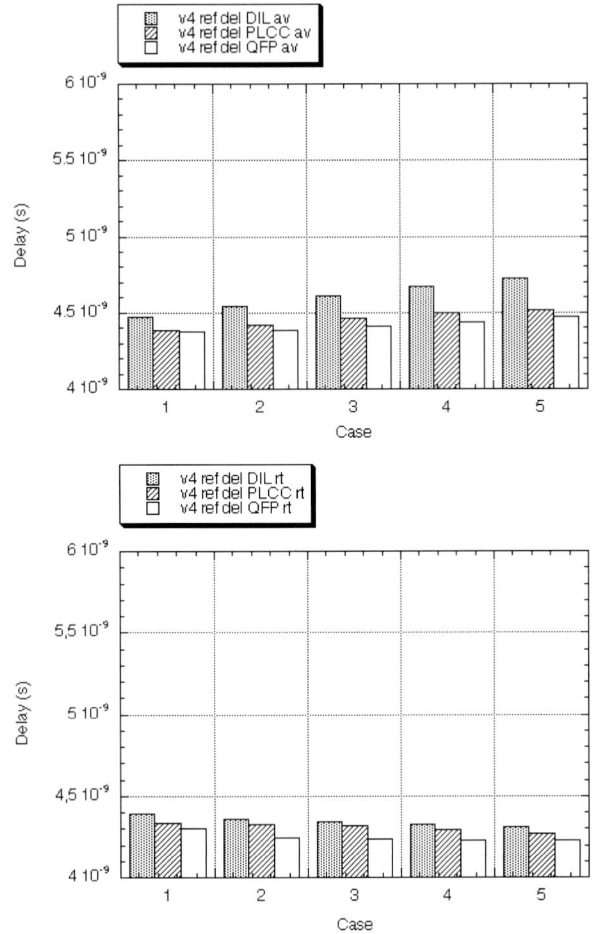

Figure 4.30. Reference delay with line 4 as victim. The lines switching in each case are: 4+5, 1+4+5, 1+2+4+5, 1+2+3+4+6, 1+2+3+4+5+6.

shows the expected behavior of advancing and retarding the transition. It is interesting to note that in the case of retarding transitions, Case 5 and 6 appear to have a smaller effect than Case 4, although there is a greater number of lines switching. The explanation is that, in the case of retarding transition, simultaneous coupling produces an important undershoot (or overshoot) because the coupled signal is opposite to the victim line transition. Although this under(over)shoot is correspondingly more significant for Cases 5 and 6, this fact does not influence the transition delay as much as other cases. On the other hand, the advancing transitions do not cause any under(over)shoot,

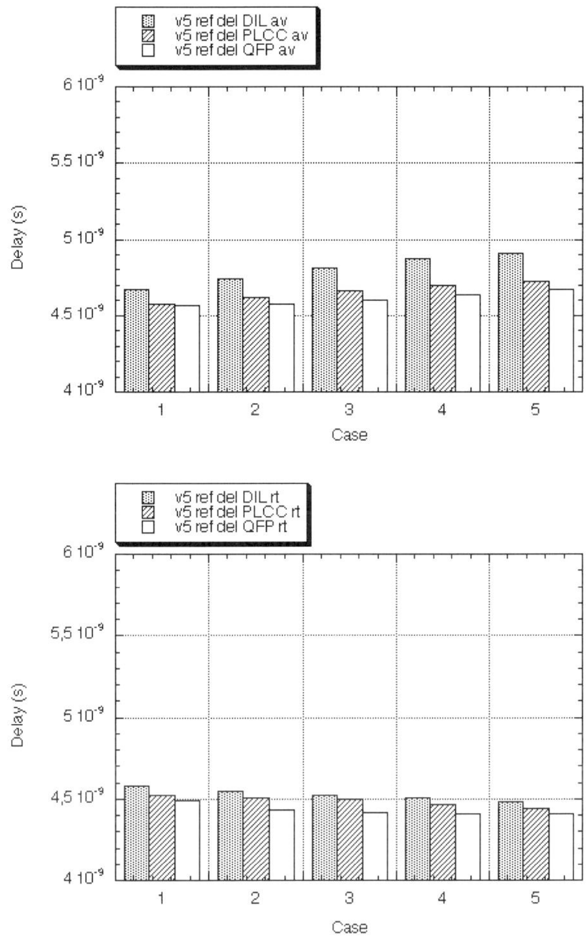

Figure 4.31. Reference delay with line 5 as victim. The lines switching in each case are: 4+5, 2+4+5, 2+4+5+6, 1+2+4+5+6, 1+2+3+4+5+6.

and therefore the increasing effect with a larger number of switching lines is observed.

These graphs also allow to evaluate the relative importance of package versus crosstalk influences for delay variation, by observing the difference between data corresponding to delay without package and data obtained including package parasitics. Again, the different behavior of advancing and retarding transitions should be noted. As was shown in figures 4.28 to 4.31, in the particular configuration of the simulated circuit, advancing transitions produce an increase in delay because of package parasitics (while crosstalk produces the opposite effect). This explains why, in the graphs corresponding to retard-

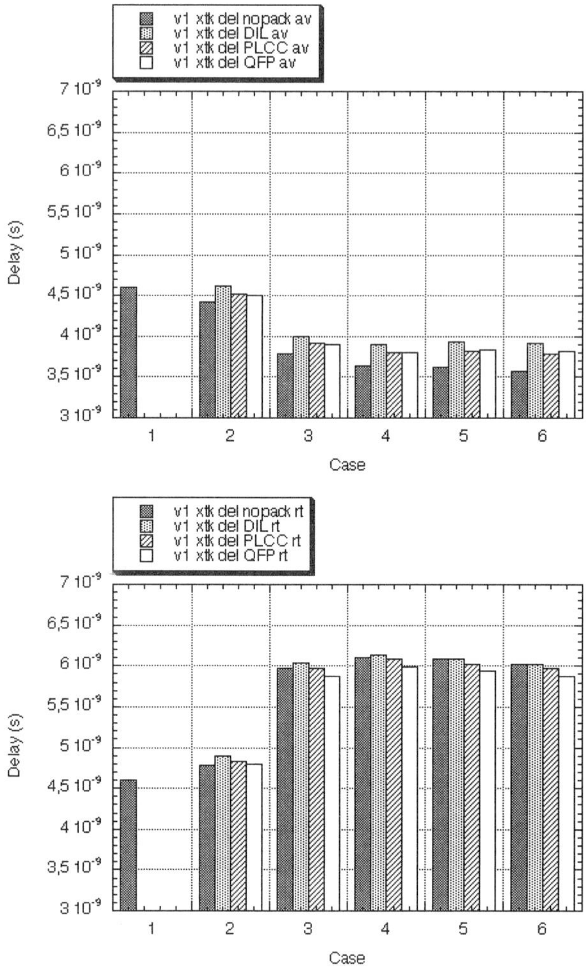

Figure 4.32. Crosstalk delay with line 1 as victim. The switching order of the lines is the same indicated in figure 4.28.

ing transitions, the delay in circuits with packages is generally smaller than the delay calculated without packages.

It can be seen from graphs that the difference between the data with and without packages is small, from 50 ps to 400 ps, while the crosstalk effect produces a much greater variation, ranging from 500 ps to 2 ns.

5.3 Discussion

The results shown in this section indicate an important dependence of delay on activity in neighboring lines. The relative importance of this effect depends

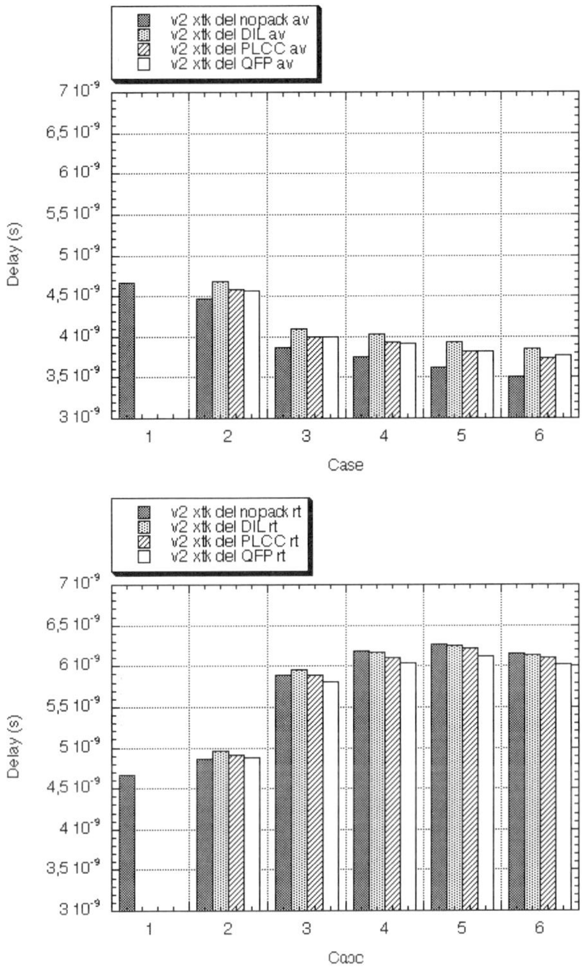

Figure 4.33. Crosstalk delay with line 2 as victim. The switching order of the lines is the same indicated in figure 4.29.

on the nominal delays involved: if these are small, as happens for example in pipelined structures, the crosstalk-induced delay can be 30% or more of the path delay.

Package parasitics have also an impact on path delay, as shown in figures 4.28 to 4.31. However, the most important influence on delay variation is due to crosstalk for closely spaced lines and the package has a relatively minor influence. This is so even though the pin-out considered (only one VDD and GND connection at opposite pins) does not favor low simultaneous switching noise. In this sense, the results presented here may be considered as a worst

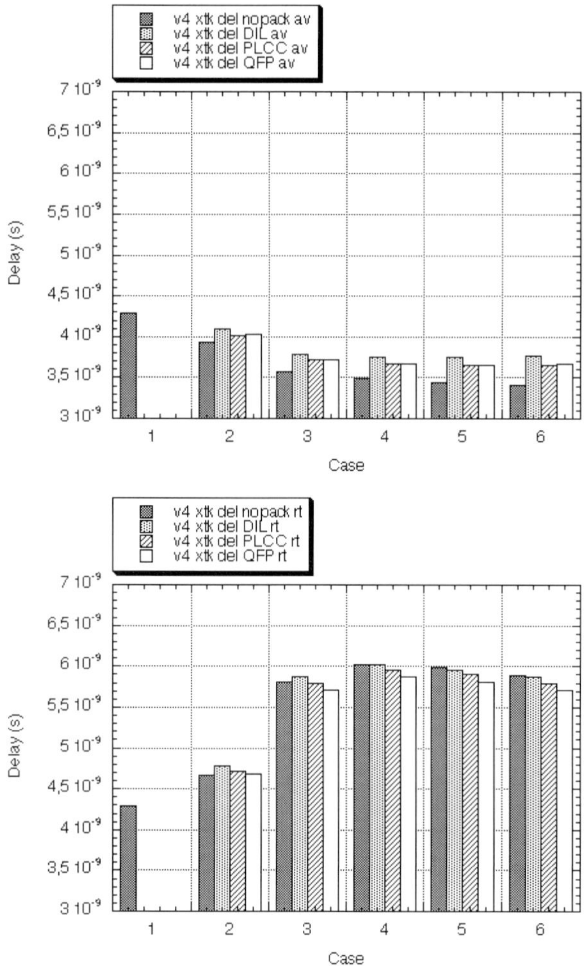

Figure 4.34. Crosstalk delay with line 4 as victim. The switching order of the lines is the same indicated in figure 4.30.

case with respect to package behavior, because several improvements, like better pin assignment and the use of on-chip or on-package decoupling capacitors are able to reduce the influence of the package [41].

References

[1] Seraphim D. P., Lasky R., and Li C.-Y. *Principles of Electronic Packaging*. McGraw–Hill, 1989.

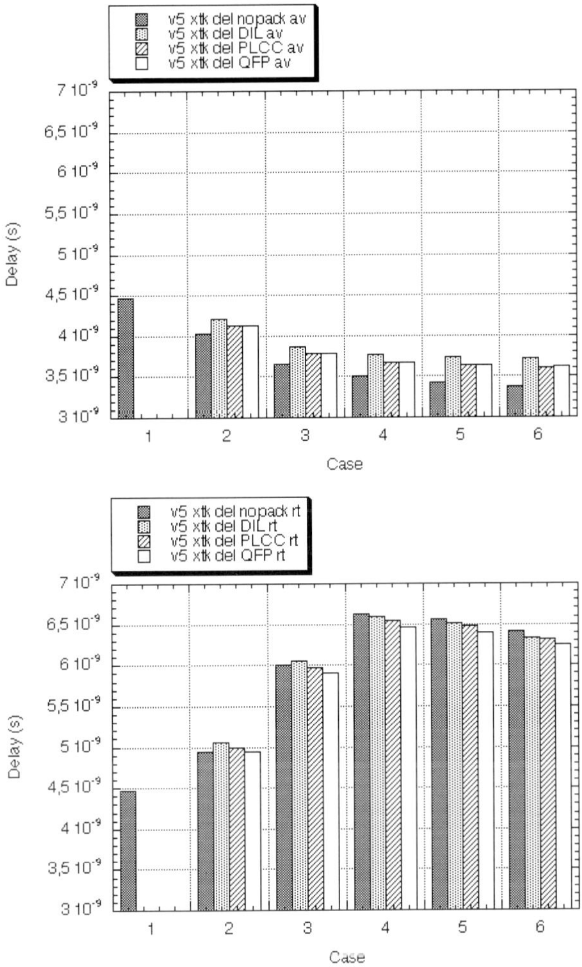

Figure 4.35. Crosstalk delay with line 5 as victim. The switching order of the lines is the same indicated in figure 4.31.

[2] Bakoglu H. B. *Circuits, Interconnections and Packaging for VLSI.* Addison-Wesley, 1990.

[3] Johnson B., Quarles T., Newton A. R., Pederson D. O., and Sangiovanni-Vicentelli A. *SPICE3 Version 3e User's Manual*. University of California, Dpt. of Electrical Engineering and Computer Sciences. University of California. Berkeley, CA, 94720, 1991.

[4] Jr. Blodgett A. J. "Microelectronic Packaging". *Scientific American*, 249, July 1983.

[5] Terasawa M., Minami S., and Rubin J. "A Comparison of Thin Film, Thick Film, and Co-Fired High Density Ceramic Multilayer with the Combined Technology: T&T HDCM

(Thin Film and Thick Film High Density Ceramic Module". *The Int. Journal for Hybrid Electronics*, 6(1), 1983.

[6] Ho C. W., Chance D. A., Bajorek C. H., and Acosta R. E. "The Thin-Film Module as a High-Performance Semiconductor Package". *IBM J. Res. and Develop.*, 26(3), May 1982.

[7] Lai Z. and Liu J. The nordic electronics packaging guideline. http://extra.ivf.se/ngl/, 2000.

[8] Brown K.M. *Physical achitecture of VLSI systems*, chapter 2, pages 23–98. John Wiley & Sons, 1994.

[9] Davidson E. E. "Electrical Design of a High Speed Computer Package". *IBM J. Res. Develop.*, 26(3), May 1982.

[10] Quint D. W., Aziz A., Kaw R., and Perezalonso F. J. "Measurement of R, L, and C Parameters in VLSI Packages". *Hewlett-Packard Journal*, October 1990.

[11] Vidano R. P., Cummings J. P., Jensen R. J., Walters W. L., and Helix M. J. "Technology and Design for High Speed Digital Components in Advanced Applications". In *33rd Electronic Components Conference*, 1983.

[12] Val C. M. and Martin J. E. "A New Chip Carrier for High Performance Applications: Integrated Decoupling Capacitor Chip Carrier (IDCC)". *IEEE Trans. on Components, Hybrids and Manufacturing Technology*, 6(3), September 1983.

[13] Katopis G. A. "Delta–I Noise Specification for a High–Performance Computing Machine". *Proceedings of the IEEE*, 73(9), September 1985.

[14] Senthinathan R. and Prince J. L. "Simultaneous Switching Ground Noise Calculation for Packaged CMOS Devices". *IEEE Journal of Solid State Circuits*, 26(11), November 1991.

[15] Ruehli A.E. "Equivalent circuit models for three-dimensional multiconductor systems". *IEEE Trans. on Microwave Theory and Techniques*, 22(3), March 1974.

[16] Ruehli A. E. "Inductance calculations in a complex integrated circuit environment". *IBM J. of Research and Development*, 16, September 1972.

[17] Ramo S., Whinnery J. R., and Van Duzer T. *Fields and Waves in Communication Electronics*. Wiley and Sons, 2nd edition, 1984.

[18] Kamon M., Tsuk M.J., and White J. "Fasthenry: a multipole-accelerated 3-D inductance extraction program". *IEEE Trans. on Microwave Theory and Techniques*, 42(9), September 1994.

[19] Ruehli A.E. and Heeb H. "Circuit models for three-dimensional geometries including dielectrics". *IEEE Trans. on Microwave Theory and Techniques*, 40(7), July 1992.

[20] Kamon M. Efficient techniques for inductance extraction of complex 3-d geometries. Master's thesis, Massachusets Institute of Technology, February 1994.

[21] Martens L. *High-frequency characterization of electronic packaging*. Kluwer Academic Publishers, 1998.

[22] Wartenberg S.A. *RF measurements of die and packages*. Artech House, 2002.

[23] Young B. *Digital signal integrity*. Prentice Hall PTR, 2001.

[24] Jong J.-M. and Tripathi V.K. "Time-domain characterization of interconnect discontinuities in high-speed circuits". *IEEE Trans. on Components, Hybrids and Manufacturing Technology*, 15(4), August 1992.

[25] Bruckstein A. and Kailath T. "An inverse scattering framework for several problems in signal processing". *IEEE ASSP Magazine*, 4(1), January 1987.

[26] Burkhart S.C. and Wilcox R.B. "Arbitrary pulse shape synthesis via nonuniform transmission lines". *IEEE Trans. on Microwave Theory and Techniques*, 38(10), October 1990.

[27] Tsai C.-T. and Yip W.-Y. "An experimental technique for full package inductance matrix characterization". *IEEE Trans. on Components, Packaging and Manufacturing Technology–Part B*, 19(2), May 1996.

[28] Young B. and Sparkman A. "Measurement of package inductance and capacitance matrices". *IEEE Trans. on Components, Packaging and Manufacturing Technology–Part B*, 19(1), February 1996.

[29] Jeong J., Nam S., Shin Y.S., Kim Y.S., and Jeong J. "Electrical characterization of Ball Grid Array packages from S-Parameter measurements below 500 MHz". *IEEE Trans. on Advanced Packaging*, 22(3), August 1999.

[30] Horng T.-S., Wu S.-M., and Shih C. "Complete methodology for electrical modeling of RFIC packages". *IEEE Trans. on Advanced Packaging*, 24(4), November 2001.

[31] ANSI/EIA-656A. I/O Buffer Information Specification, 1999.

[32] Vakanas L. P., Cangellaris A. C., and Palusinski O. A. "Scattering Parameter-Based Simulation of Transients in Lossy Nonlinearly Terminated Packaging Interconnections". *IEEE Trans. on Components, Packaging and Manufacturing Technology–Part B*, 17(4), November 1994.

[33] Achar R. and nakhla M.S. "Efficient transient simulation of embedded subnetworks characterized by S-parameters in the presence of nonlinear elements". *IEEE Trans. on Microwave Theory and Techniques*, 46(12), December 1998.

[34] Ruehli A.E. and Cangellaris A.C. "Progress in the methodologies for the electrical modeling of interconnects and electronic packages". *Proceedings of the IEEE*, 89(5), May 2001.

[35] Yee K.S. "Numerical solution of initial boundary value problems involving Maxwell's equations in isotropic media". *IEEE Trans. on Antennas and Propagation*, AP-14:302–307, May 1966.

[36] Moll F. and Fang J. "Extraction of equivalent circuit models of package power supply distribution systems from full wave electromagnetic simulations". In *7th Topical Meeting on Electrical Performance of Electronic Packaging*, October 1998.

[37] Sigrity Inc. *SPEED Manual*, http:/www.sigrity.com edition, 2001.

[38] Avant! Corp. *Star-Hspice 2001.2*, June 2001.

[39] Moll F. and Roca M. "Effect of package parasitics and crosstalk on signal delay". In *Signal Propagation in Interconnects*, 2001.

[40] Moll F., Roca M., Rubio A., and Sicard E. "Analysis and Measurement of Crosstalk-Induced Delay in Integrated Circuits". In *Signal Propagation in Interconnects*. Kluwer, 1998.

[41] Garben B., McAllister M.F., Becker W.D., and Frech R. "Mid-Frequency Delta-I Noise Analysis of Complex Computer System Boards with Multiprocessor Modules and Verification by Measurements". *IEEE Trans. on Advanced Packaging*, 24(3), August 2001.

Chapter 5

TECHNIQUES FOR AVOIDING INTERCONNECTION NOISE

The previous chapters were concerned about the correct modeling of interconnection-related problems: how to model the phenomena in terms of electric circuits and a description of these phenomena. It was shown that, under certain circumstances, interconnections cause signal perturbations which can negatively affect overall performance, maybe causing transient or permanent errors in a digital system.

This chapter concentrates in analyzing the circumstances causing such problems, and consequently, giving a set of intuitive rules so that the effect of interconnections is minimized. In most cases, a detailed analysis using advanced tools is necessary, but it is very useful to have, at least, an intuitive perception of the reason why the tools do their job so an attempt can be made not to incur problems from the very beginning of the design process.

1. Crosstalk avoidance

Based on the description of chapter 3, the there are four main parameters influencing the crosstalk problem, ordered from physical to design level:

1 Technology. The materials of lines and dielectrics and their minimum sizes and distances determine the worst coupling case. Manufacturing tolerances also must be considered when analyzing the problem.

2 Interconnection layout. By choosing appropriate line dimensions and spacings, crosstalk magnitude can be controlled.

3 Driver sizing. Also, the relative characteristics of drivers in coupled lines have an important influence on the effect the signals produce.

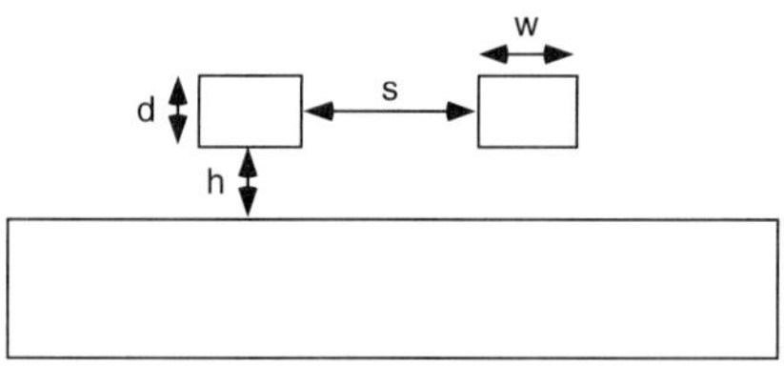

Figure 5.1. Two lines in the same metal level on a silicon grounded substrate.

4 Tolerant circuits. Finally, if the presence of crosstalk noise cannot be prevented by the previous parameters, the design of circuits which are not affected by these perturbations can be considered.

We will give some examples showing the influence of the respective parameters on the crosstalk problem.

1.1 Technology solutions

Considering the capacitive crosstalk models in chapter 3, the manufacturing process will influence the line resistance and capacitance (both coupling capacitance and capacitance to ground). The main aspect related to technology consists of the ratio of vertical to horizontal dimensions.

1.1.1 Vertical dimensions

Generally speaking, the choice of horizontal dimensions (width of lines, and length) above minimum values determined by the technology is the designer's responsibility, while all vertical dimensions depend exclusively on the chosen manufacturing process.

A very simplified but insightful analysis can be made of the influence of vertical and horizontal dimensions of lines on the crosstalk problem. We consider as a simple example the case of two lines placed over a silicon substrate as depicted in figure 5.1. A first approach to the parasitic ground capacitance per line (C_{line}) and coupling capacitance between the two lines ($C_{coupling}$) can be obtained using the infinite parallel plane approximation:

$$C_{line} \approx \epsilon_{ox} \frac{lw}{h} \tag{5.1}$$

$$C_{coupling} \approx \epsilon_{ox} \frac{dl}{s} \tag{5.2}$$

If we consider the simple ideal model of lumped capacitances described in chapter 3, the amplitude of a crosstalk spurious signal can be written as:

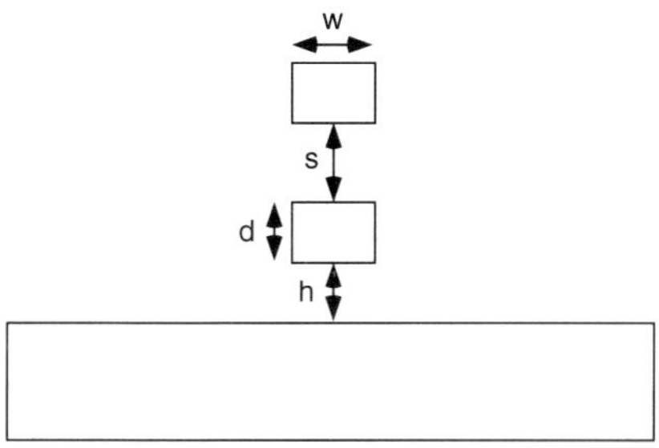

Figure 5.2. Two overlapping lines on a silicon grounded substrate.

$$\Delta V \approx \frac{V_{DD}}{1 + \frac{C_{line}}{C_{coupling}}} \tag{5.3}$$

Therefore, in order to minimize noise, the ratio between C_{line} and $C_{coupling}$ must be at a maximum. This ratio is expressed as:

$$\frac{C_{line}}{C_{coupling}} \approx \frac{ws}{dh} \tag{5.4}$$

From the previous expression, it is seen that both d and h (the vertical dimensions) must be as small as possible to minimize crosstalk. However, it must be taken into account that a reduction in h implies an increase in ground capacitance (hence the crosstalk reduction) and this causes an increase in terms of the line's signal propagation delay. Also, a reduction in d causes an increase in the line's resistance and, therefore, also has an impact on propagation delay. In this respect, the use of copper lines with a lower resistivity than aluminum allows the use of thinner lines and therefore reduces crosstalk. However, as line pitch (s) becomes smaller, coupling capacitance is more and more dependent on fringing capacitance effects (in this case, on line width), so reducing d does not translate into a proportional reduction in $C_{coupling}$.

Likewise, a similar analysis for crosstalk between lines in different layers (figure 5.2) leads to the conclusion that to minimize crosstalk, the interlayer dielectric (s) should be as thick as possible. Line thickness (d) has a second-order influence through fringing capacitance, but the effect is the same as in the previous case: a smaller d gives a smaller crosstalk effect.

In summary, the vertical dimensions could be tuned to reduce the effect of crosstalk, but at the expense of an increase in propagation delay which, on the other hand, can be compensated by driver sizing or layout techniques, translating the problem to an increase in power or area.

1.2 Interconnection layout

As pointed out in section 1.1.1, the crosstalk effect also depends on horizontal dimensions of the lines, defined by the layout. Layout techniques to minimize crosstalk can be implemented in automatic place and route tools to drive the physical design stage of the circuit [1].

Two different changes in the layout interconnects are principally analyzed here:

- changes in line width
- screening techniques and changes in distance between lines.

1.2.1 Increase in line width

In order to illustrate the impact of change on line width, let us consider two parallel coupled lines, one affecting line and one victim line. As shown in expression 5.4, a possible solution to reduce the noise signal is to increase the width of both coupled lines, as shown with a HSPICE simulation in figure 5.3. On the top panel, the spurious signal produced in the victim line is shown. It can be seen that the amplitude is reduced when the width is increased. This is due to the increase in ground capacitances while coupling capacitances remain approximately the same because the distance between lines does not vary. As can be seen in the bottom line, the signal propagation on the affecting line will slow down due to the increase in ground capacitance. There is a trade-off between crosstalk noise and propagation delay. In order to reduce the impact on delay and on area overhead, the minimum increase in width so that crosstalk amplitude is below the required dynamic noise margin is considered as the best solution.

1.2.2 Screening and line separation

The second technique presented in this section is based on screening effects. Screening techniques consist in adding a close line connected to ground or V_{DD}, either between the coupled lines, or above them. In both variants the capacitance to ground of each line increases and therefore crosstalk decreases. If the screen is inserted between the lines, the coupling capacitance is also very much reduced, adding to the crosstalk reduction effect.

In order to illustrate the importance of screening in terms of crosstalk noise immunity, we can observe figure 5.4. In this figure two curves are depicted, representing coupling capacitance between lines in function of line separation. One of the curves (labeled NON-SCREEN EFFECT) is the coupling capacitance of the two lines. The other curve represents the coupling capacitance of both lines at the same distance as the previous curve when n grounded lines are inserted between them (with n from 1 to 5). It can be seen that adding multiple

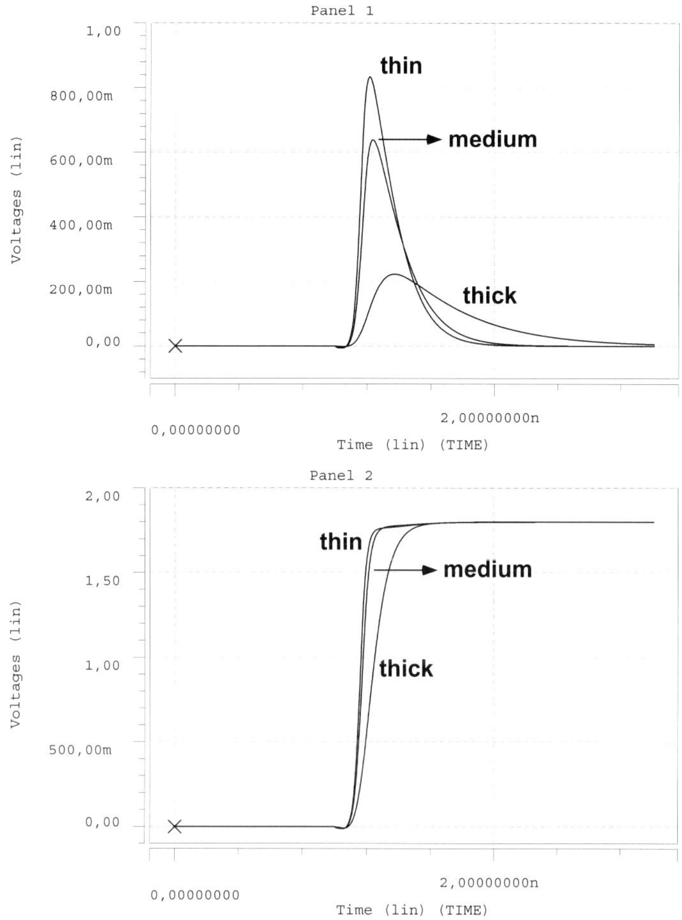

Figure 5.3. Simulation results for different width of lines.

screen lines implies a very high reduction in coupling capacitance (represented in log scale) and this reduction is much more important than the reduction in the non-screening distribution when only increase in distance between lines is considered.

It should be noted that the grounded screen line must be noise-free in order to be effective. As they are connected to one of the power lines, the existence of switching noise in the power distribution network tends to reduce the advantages of screening lines.

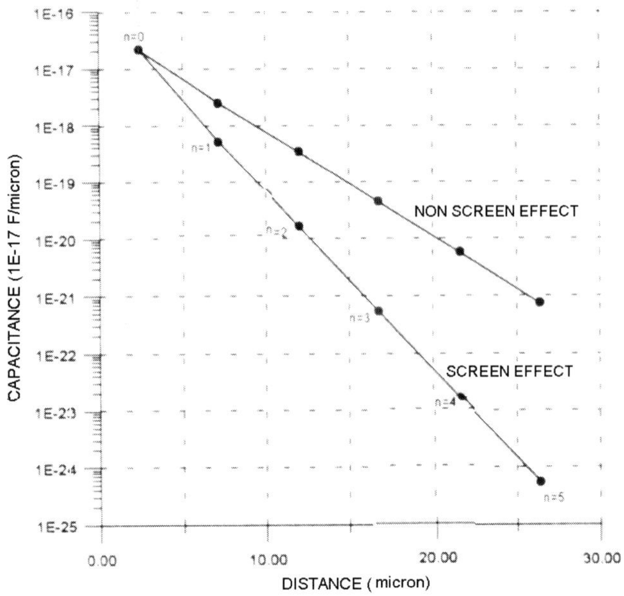

Figure 5.4. Reduction of coupling capacitance due to screen effect.

1.2.3 Analysis of crosstalk avoidance layout rules

Presented here as an example, a simple analysis of the efficiency of different rules related to interconnect layout solutions for avoiding or minimizing crosstalk depending on the cross-section dimensions of the lines.

The aspect ratio (height to width) of the line's cross-section develops along with technology. As width shrinks with each generation of technology according to Moore's law, the line's height is also reduced, but on a scale smaller than width in order to not increase line resistance very much. Therefore, aspect ratio (A/R) tends to increase with each generation. This is shown in table 5.1, where the predictions of the International Technology Roadmap for Semiconductors (ITRS) for wiring are displayed [2].

Based on this table, we will analyze the effectiveness of different schemes for minimizing crosstalk depending on different values of the aspect ratio and vertical dimensions. Let us consider the structures presented in figure 5.5, where the α parameter represents aspect ratio (A/R). The first structure is composed of two coupled lines on a silicon substrate in a typical 0.18 μm technology. This structure shows an α parameter equal to 1.6, and it will be used as an example of a modern technology (year 2001 according to table 5.1) with minimum size lines, typical of local wiring. The second case considers a smaller value of α (0.4) and it may represent a modern technology where the lines are

year	*2001*	*2005*	*2010*
Local wiring pitch (nm)	350	185	105
Local wiring A/R	1.6	1.7	1.8
Interm. wiring pitch (nm)	450	240	135
Interm. wiring A/R	1.6	1.7	1.8
Global wiring pitch (nm)	670	360	205
Global wiring A/R	2.0	2.2	2.3

Table 5.1. ITRS 2001 predictions for wiring dimensions [2].

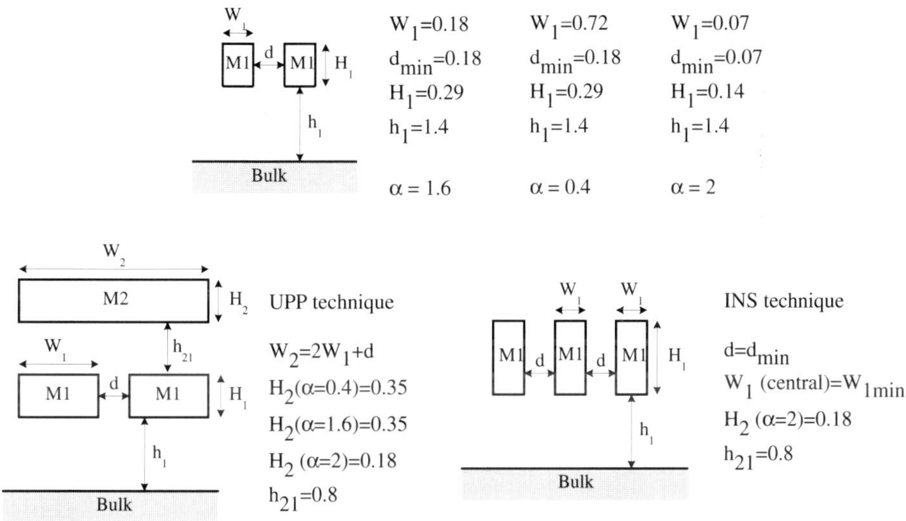

Figure 5.5. Physical structure of the cases considered for the evaluation of crosstalk avoidance techniques. All dimensions expressed in μm.

wider than the minimum value, typical of longer interconnection, for example a bus. Finally, the third case represents a situation where α is equal to 2. It corresponds to an extrapolation of current cross-section dimensions for possible future nanometric technology with a wiring pitch of 140 nm.

Interlayer dielectric thickness does not scale as aggressively as metal thickness. For this reason, we consider a constant dielectric thickness for the three structures analyzed. The derivation of the parameters characterizing the behavior of the coupled lines has been done using a 2D electromagnetic field solver integrated in HSPICE [3]. From the cross-section geometry description of the lines, the field solver provides the RLC matrix description of the coupling. A transmission line model is considered in the simulation environment.

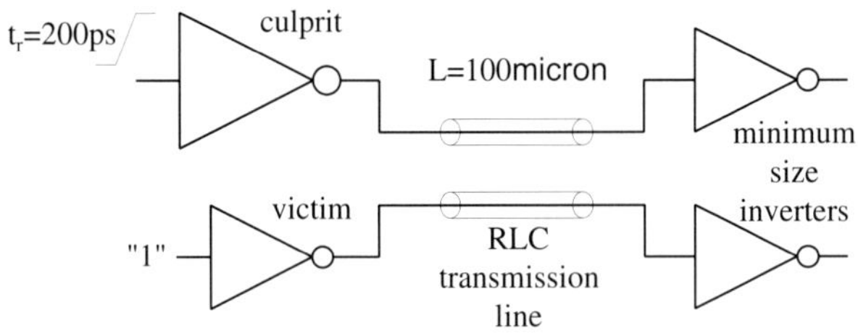

Figure 5.6. Schematic of the circuit used to analyze crosstalk.

The circuit configuration used can be seen in figure 5.6. Two CMOS inverters drive two coupled lines of length equal to 100 μm, which are feeding two other minimum size inverters. The culprit line inverter is dimensioned to have 10 times more driving capability than the victim line inverter. An input transition with $t_r = 200\,\text{ps}$ is applied to the input of the culprit inverter, whereas the input of the victim inverter is held to a logic value of one (V_{DD} voltage). A spurious signal is expected to appear on the victim line due to the coupling. The amplitude of this spurious signal, ΔV, will be used to quantify the crosstalk effect.

Three strategies (which can be seen also in figure 5.5) to reduce crosstalk will be simulated and compared, that is:

SPA increasing spacing between lines to a multiple of the minimum distance, d_{min}.

UPP adding a screen ground line in an immediate upper metal layer.

INS inserting a screen ground line in between lines.

Some HSPICE simulations have been conducted for this circuit and the 3 line topologies of figure 5.5, using a level 49 model for the MOSFET devices, and a supply voltage of $V_{DD} = 1.8\,\text{V}$. The results obtained can be summarized as follows:

- Case α equal to 1.6 (Today's technology). Figure 5.7 shows an HSPICE simulations for the two coupled lines geometry described before. Along with the original two coupled lines (a), results are shown for three crosstalk minimization techniques: (b) spacing out the lines by $2 \cdot d_{min}$ (SPA), (c) adding a ground line in the immediate upper metal layer (UPP), and (d) combining both techniques. Figure 5.8 shows a HSPICE simulations, comparing: (a) the original two coupled lines, (b) with spacing lines by $2 \cdot d_{min}$, (c) spacing the lines by $3 \cdot d_{min}$, and (d) inserting an screen line (minimum width) between them. This last case involves the same increase in area as (c) spacing solution.

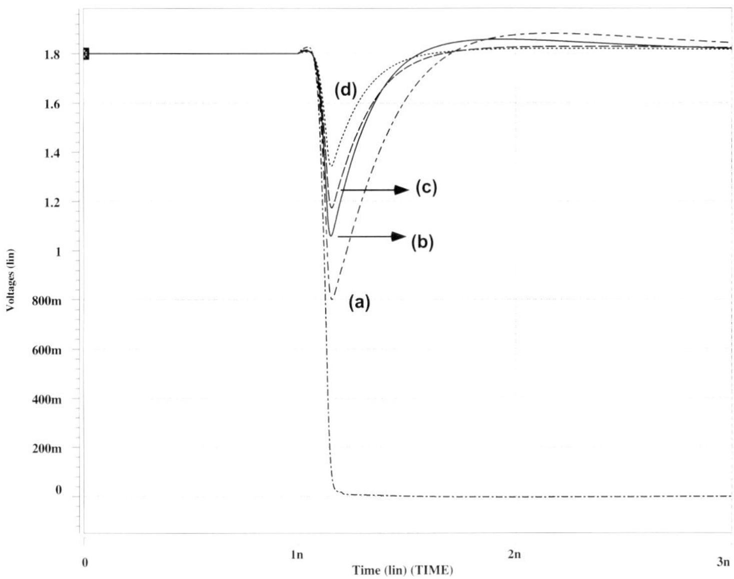

Figure 5.7. Crosstalk spurious signal for the case $\alpha = 1.6$ (a) minimal space, (b) SPA ($2 \cdot d_{min}$), (c) UPP, (d) SPA+UPP techniques.

Considering results from Figure 5.7 the results obtained with the SPA technique can be seen as the UPP technique improves and the consideration of both techniques simultaneously involves a reduction of nearly 60 % of the glitch amplitude. Observing Figure 5.8, it can be concluded than the INS technique is the best solution, minimizing the crosstalk effect (the glitch practically disappears). It is clear that SPA ($3 \cdot d_{min}$) and INS presents the same area overhead, but, on the other hand, INS improves the SPA results a great deal.

- Case α equal to 0.4. (A bus example in a modern technology). Figure 5.9 shows a HSPICE simulations for the two coupled lines in the structure corresponding to $\alpha = 0.4$. Results are shown for (a) the two original coupled lines, (b) spacing out the lines by $2 \cdot d_{min}$ (SPA), (c) adding a ground line in the immediate upper metal layer (UPP) and (d) combining both techniques.

  Figure 5.10 shows a HSPICE simulations, comparing: (a) the original two coupled lines, (b) lines spaced twice the minimum distance, (c) spacing the lines by $3 \cdot d_{min}$, and (d) inserting a screen line (minimum width) between them. As before, this latter case implies the same increase in area as the (c) spacing solution.

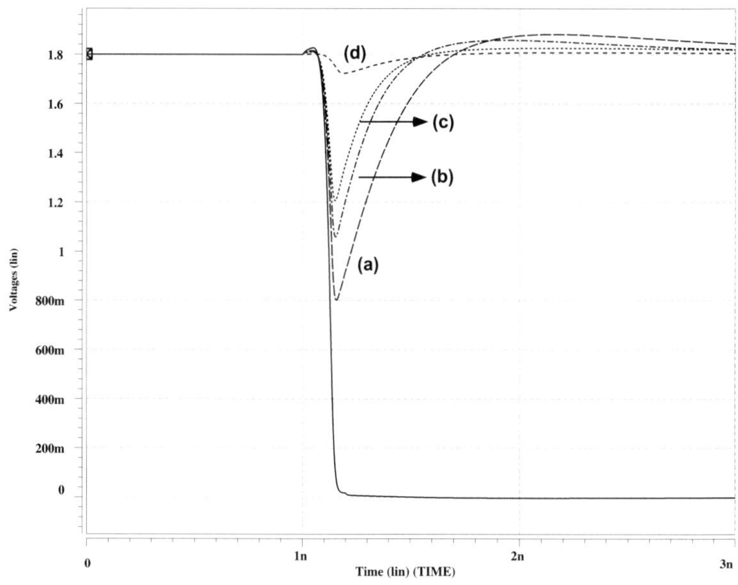

Figure 5.8. Crosstalk spurious signal for the case $\alpha = 1.6$ (a) minimal space, (b) SPA ($2 \cdot d_{min}$), (c) SPA ($3 \cdot d_{min}$), (d) INS technique.

For the case of buses the results are different. Considering the results from Figure 5.9, SPA technique now seems to be better than UPP technique and considering both techniques simultaneously improves the results. Observing Figure 5.10, the same conclusions as in the previous case are obtained.

- Case α equal to 2. (Nanometric advanced technology). Figure 5.11 presents HSPICE simulations for the two coupled lines geometries depicted corresponding to the structure with $\alpha = 2$. The simulations correspond to (a) the two original coupled lines, (b) spacing out the lines by $2 \cdot d_{min}$, (c) adding a ground line in the immediate upper metal layer (UPP), and (d) combining both techniques. Figure 5.12 shows HSPICE simulations, comparing: (a) the original two coupled lines, (b) lines spaced at twice the minimum distance, (c) spacing the lines by $3 \cdot d_{min}$, and (d) inserting a screen line (minimum width) between them.

  Considering the results from Figure 5.11 it can be seen that the UPP technique does not have any effect as a technique for preventing crosstalk. SPA by $2 \cdot d_{min}$ technique allows crosstalk reduction but with a smaller effect than in the previous cases. Observing Figure 5.12, it can be concluded that INS technique is the best solution, minimizing the crosstalk effect (the glitch practically disappears).

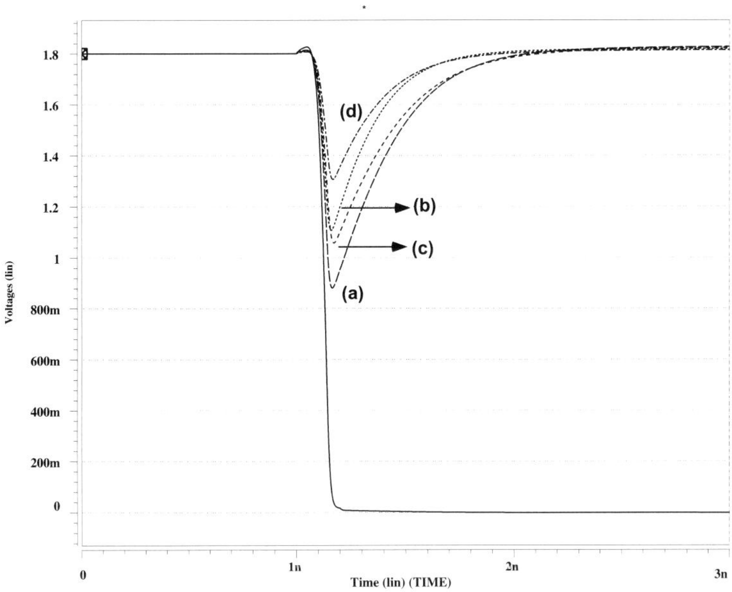

Figure 5.9. Crosstalk spurious signal for the case $\alpha = 0.4$ (a) minimal space, (b) SPA ($2 \cdot d_{min}$), (c) UPP, (d) SPA+UPP techniques.

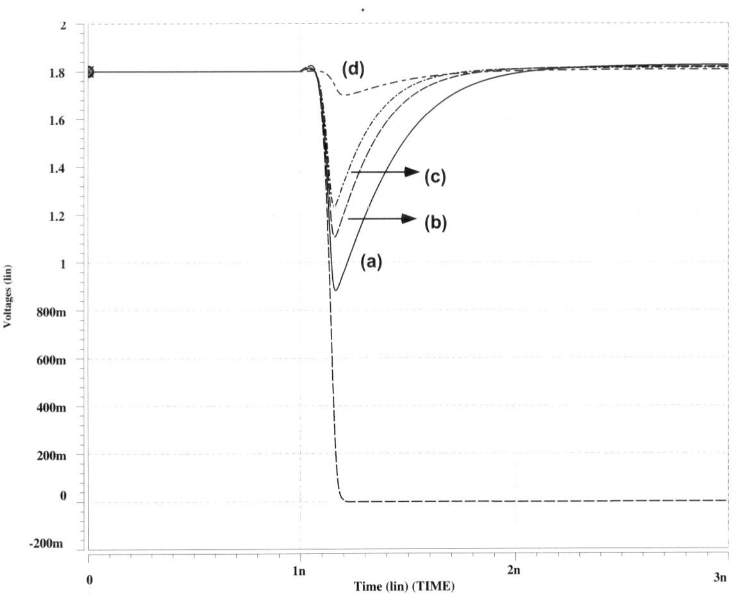

Figure 5.10. Crosstalk spurious signal for the case $\alpha = 0.4$ (a) minimal space, (b) SPA ($2 \cdot d_{min}$), (c) SPA ($3 \cdot d_{min}$), (d) INS technique.

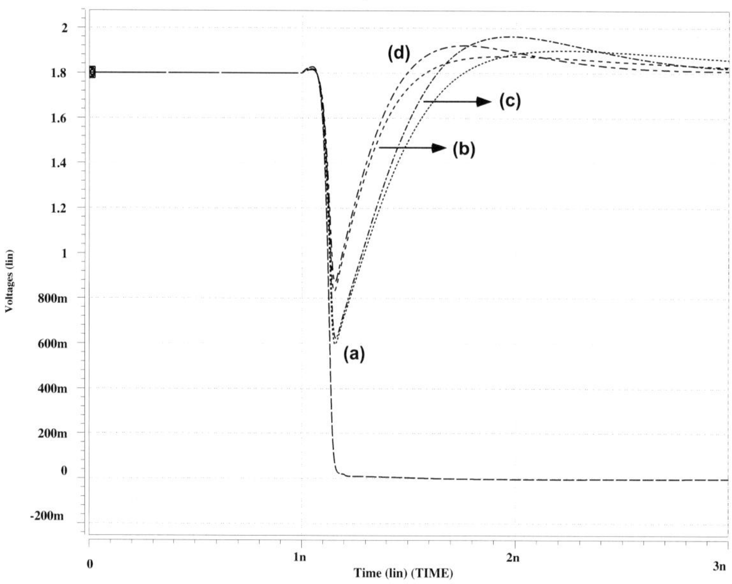

Figure 5.11. Crosstalk spurious signal for the case $\alpha = 2$ (a) minimal space, (b) SPA ($2 \cdot d_{min}$), (c) UPP, (d) SPA+UPP.

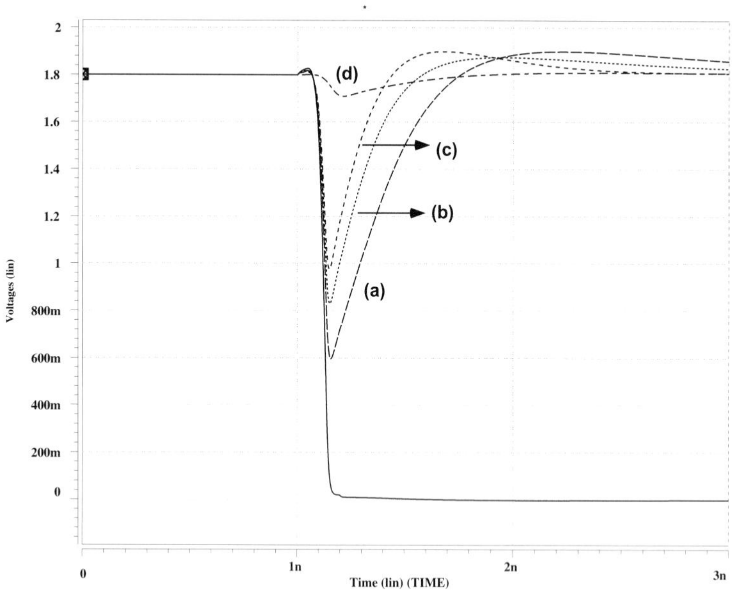

Figure 5.12. Crosstalk spurious signal for the case $\alpha = 2$ (a) minimal space, (b) SPA ($2 \cdot d_{min}$), (c) SPA ($3 \cdot d_{min}$), (d) INS.

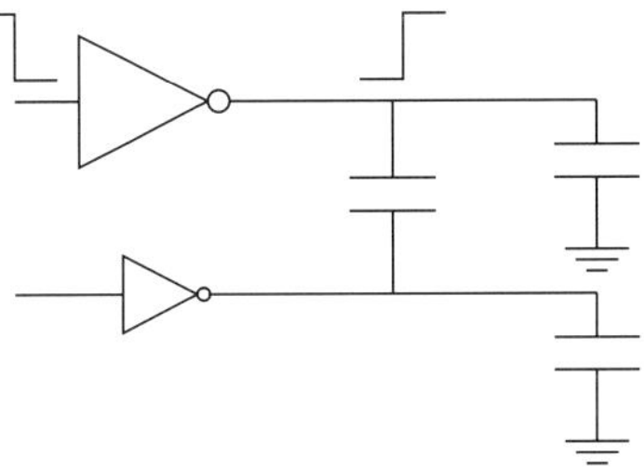

Figure 5.13. Circuit for driver size analysis.

From these results, it is clear that different alternatives for reducing crosstalk effects should be applied depending on the value of α. In today's technologies, and when minimum width lines are concerned, the combination of spacing out and screening an immediate upper metal layer with a ground line is better. If wider lines are involved, then spacing out is a good solution. Finally, it may be predicted that if current reduction trend is maintained in future technologies, then the solution of introducing a screening ground line in between the two coupled lines will be the best option, both in terms of crosstalk minimization and area overhead.

1.3 Driver sizing

As was analyzed in section 1.1 of chapter 3, the optimum crosstalk noise amplitude corresponds to a ratio of driver resistances conducting these lines equal to one. This result can be considered as a design rule for preventing crosstalk.

A demonstration of this rule can be made by considering the circuit in figure 5.13 made up of two lines modeled as a capacitance to ground and a coupling capacitance between them. Each line is driven by a CMOS inverter. We have analyzed the effect of crosstalk by electrical simulation (using HSPICE), making a comparison between the cases when the ratio between transistor sizes of the affected and affecting line inverters are 1:1, 10:1 and 100:1. These simulation results are presented in figure 5.14. In the upper panel the expected result that the amplitude and width of the crosstalk spurious signal is reduced when the drivers' ratio is 1:1 and it is at its maximum when the drivers' ratio is 1:100 can be seen. Meanwhile, the affecting signal is represented in the lower panel of figure 5.14, and it can be seen that the propagation time increases when crosstalk reduces. In order to reduce this penalty in propagation signal delay, a rescaling of the two drivers may be necessary to obtain acceptable propagation times.

Another technique related to driver sizing is the use of buffers or repeaters along the line to minimize crosstalk [4]. This technique is based on the re-

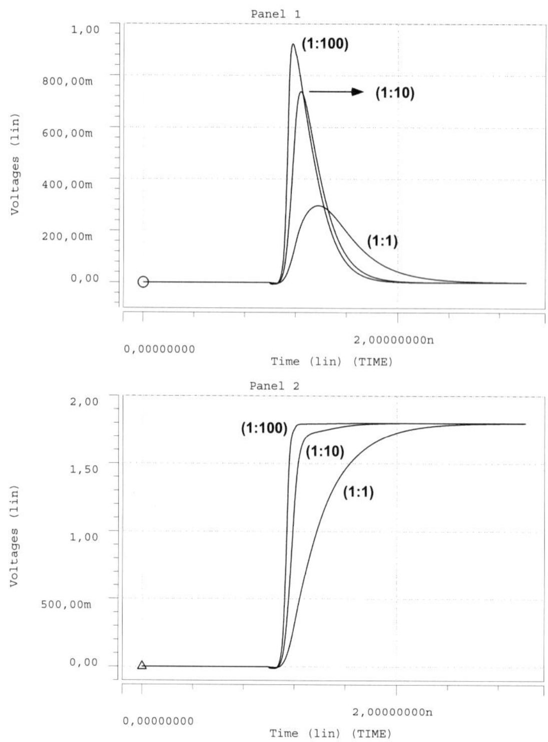

Figure 5.14. Simulation results of driver sizing analysis.

duction of the RC constant of driver-line. By using the same driver repeated along the line, each stage drives a smaller total capacitance (self- plus coupling capacitance) and this produces a smaller crosstalk amplitude and width, even when the capacitance ratio is the same. This was also seen in the crosstalk analysis as a function of line length of chapter 3 (figure 3.10).

1.4 Tolerant circuits

In this section we will mention two different kind of solutions. The first one is the use of differential signaling inside the chip and the second solution is the use of crosstalk-tolerant logic blocks in the parts of the circuit more susceptible to crosstalk noise.

Differential signaling methodologies are more robust to crosstalk because crosstalk will affect only one of the signals, or both signals as a common mode noise that is automatically eliminated. As a drawback, there is a complexity and area overhead compared with conventional single end signaling because two lines are needed per signal. On the other hand, due to the complementary

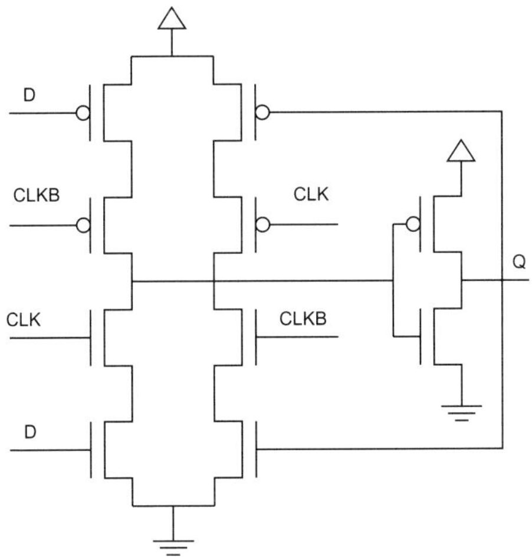

Figure 5.15. Conventional latch structure.

signaling, these technologies do not need the use of inverters in the design. It will be seen later that differential signaling is also a very good solution for reducing simultaneous switching noise.

The second solution deals with the use of tolerant blocks. Assuming that crosstalk cannot be eliminated by using technology, layout or driver techniques, the objective is to modify logic blocks so that they are tolerant to noise in some or all of their inputs. These modified blocks can be used in critical areas where noise is found to be more important.

As an example [5], a crosstalk tolerant latch is shown here. Figure 5.15 shows a conventional transistor structure for a D-latch cell. The circuit is composed by ten MOS transistors. The data line is labeled as *D*, the differential clock signals are *CLK* and *CLKB*, while *Q* is the output of the latch. The behavior of the circuit is as follows:

- when *CLK* is at 1, output Q follows data D (transparent state).
- when *CLK* is at 0, output Q maintains the same value (memory or latch state).

This conventional structure is not crosstalk tolerant to common mode noise in the *CLK/CLKB* signals. As both signals are complements of one another, a positive pulse in both of them will only affect the one which is at 0, and a negative pulse will only affect the signal which is at logic 1. While a crosstalk noise in the transparent state will only cause a transient signal at the output, if the noise occurs while the latch is in the memory state, it can change the stored value. For example, this can happen if the stored value is a logic 0, and

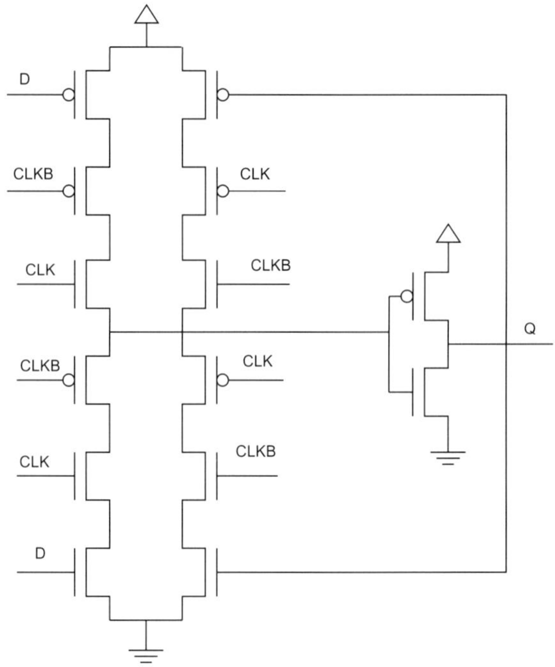

Figure 5.16. Tolerant latch structure.

a positive spurious signal arrives at both *CLK* and *CLKB* while $CLK = 0$. If this happens while $D = 1$, for a short time the latch becomes transparent and a wrong logic 1 is stored at the output. Other situations exist causing a logic error. In order to prevent these errors, a new structure could be considered.

In figure 5.16 a modified latch structure with four more transistors than the previous conventional structure (this is the penalty in block area compared to the conventional one) can be seen. This new structure is crosstalk tolerant to noise in the *CLK* lines. In figure 5.17 measurements on a tolerant structure implemented in a VLSI circuit are presented. It is shown in this figure that spurious pulses in *CLK* signal (those periods of time when *CLK* and *CLKB* are not inverted) do not affect the behavior of the latch. The behavior of this latch is the same as the behavior of the previous non-tolerant structure. This tolerant structure can be adopted as a latch in the situations on the VLSI circuit more susceptible to crosstalk noise, allowing noise immunity with a small impact in area overhead.

2. Switching noise avoidance

At package level, measures for reducing noise are strongly related to the problem of Simultaneous Switching Noise (SSN). As pointed out in chapter 1,

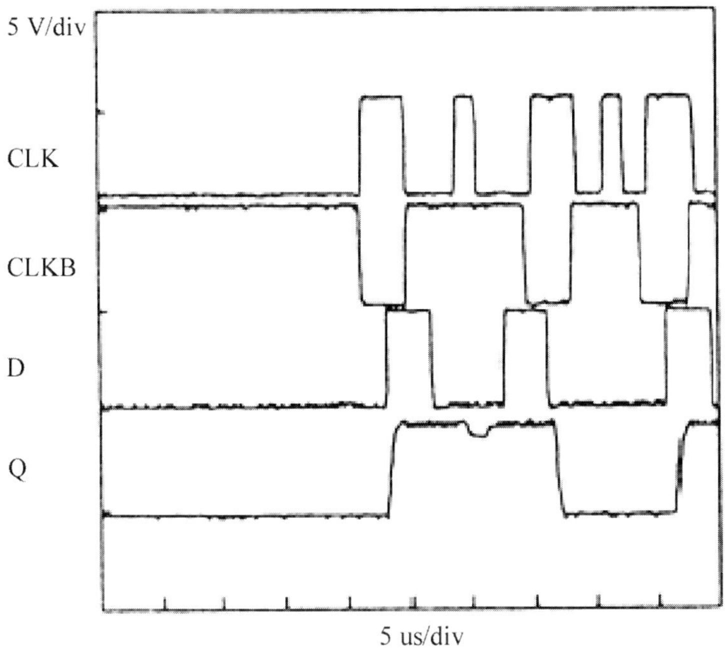

Figure 5.17. Measurements on the tolerant latch structure.

this phenomenon is directly related to the impedance of the power/ground distribution network in function of frequency: if the impedance was zero for the whole frequency range, there would not be any noise generated. The following parameters influence noise generation and are therefore candidates for avoidance rules:

1 Technology. The packaging technology determines its primary characteristics, especially with respect to the existence of power and ground planes, number of interconnection layers, etc.

2 Power and ground pin assignment. The pin assignment to physical package leads implies that the coupling between critical signals can be increased or decreased.

3 Use of decoupling capacitors. Decoupling capacitors can alleviate problems related to SSN, if they are appropriately placed. Capacitor placement is an important decision.

4 Circuit techniques. Switching circuits cause SSN, and therefore the noise generated can be controlled by controlling the way that different parts of the circuit switch.

A more detailed discussion follows, developing the different parameters.

2.1 Package technology

The impedance of the power distribution network depends on the resulting inductance, capacitance and resistance resulting from the physical structure. As explained in chapter 4, inductance depends on the current path and not on the materials used. The case is different for capacitance and resistance, directly related to the dielectric and conductive materials respectively.

At DC, the impedance is determined by the resistance. The use of low resistivity conductors, like copper or gold, gives a reduction in impedance.

However, as frequency increases, the equivalent impedance is dominated by the inductance and its value is determined by the number of connections to V_{DD} or V_{CC} and by the distance between conductors carrying current to and from the power supply. The shorter the distance between conductors (and the shorter their length), the smaller the value of the inductance and therefore the less noise generated. As a consequence, the use of small size packages has an added benefit for high integration at PCB level: that is, better electrical performance because the inductance is smaller. An example of this trend is the so-called Chip Scale Packaging (CSP) [6], in which the package is roughly the same size as the chip. This packaging strategy normally uses flip-chip type connections between chip and package to achieve the ratio between chip and package area close to unity. For larger packages, power distribution can be made through planes, so that there are several conductor layers, complicating the package structure and increasing its cost.

On the other hand, the parasitic capacitance between conductors connected to V_{DD} and conductors connected to V_{CC} causes, in conjunction with the inductance, several resonance peaks at selected frequencies. For those frequencies, the value of the impedance may rise significantly, even by an order of magnitude, or more in some cases. If the resonance frequency is close to the frequency of the switching current, then the noise problem becomes very serious. The value of capacitance can be modified using a different dielectric, so there is a technological solution for changing the resonance frequency. In practice, solutions based on design (lead assignment, addition of decoupling capacitors and circuit techniques) are easier and cheaper to implement for changing the resonance frequencies.

2.2 Use of decoupling capacitors

One way of reducing switching noise is to use decoupling capacitors. When a capacitor is placed physically close to the switching element (a package at PCB level, a chip at MCM level, or a gate or logic block at chip level), the current demanded by the switching event is partially given by the capacitor instead of the power distribution network. Therefore, the noise generated is reduced compared to the case in which all the current has to traverse the en-

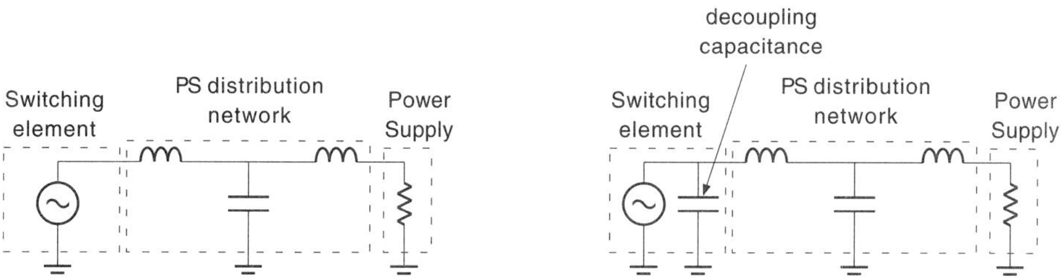

Figure 5.18. One-stage LC circuit model of a package power distribution network, with and without decoupling capacitance.

tire impedance of the distribution network from the power supply. Therefore, the design of a system's power distribution network includes the placement of different decoupling capacitors at the different hierarchical interconnection levels: motherboard, board, package, and chip.

The effect is very well illustrated by the computation of the equivalent impedance of the power distribution network seen from the switching element. We will analyze the difference between this impedance with and without decoupling capacitance at package interconnection level.

2.2.1 Switching of internal nodes

First, the case of internal switching is analyzed (see figure 4.19 in chapter 4). The equivalent impedance from the terminals of the switching element can be computed with an AC analysis of the circuit model, as shown in figure 5.18. It can be seen intuitively by examining the circuit topology of this diagram that if the decoupling capacitance is not present, the equivalent impedance rises with frequency (after one or more resonance peaks, depending on the complexity of the model). The effect of the decoupling capacitance is to short the power terminals at high frequency, and the resulting impedance is therefore smaller for high frequencies.

The effectiveness of this reduction depends on several factors:

1 The decoupling capacitor must be located as close as possible to the switching element. If possible, the best solution is to integrate on-chip capacitors for this purpose.

2 The capacitor must have a low series inductance, because otherwise the total impedance again increases at high frequencies, beyond its self-resonant frequency.

3 The inclusion of the decoupling capacitance may introduce extra resonant frequencies, which should differ from the fundamental operating frequencies of the switching current to prevent problems.

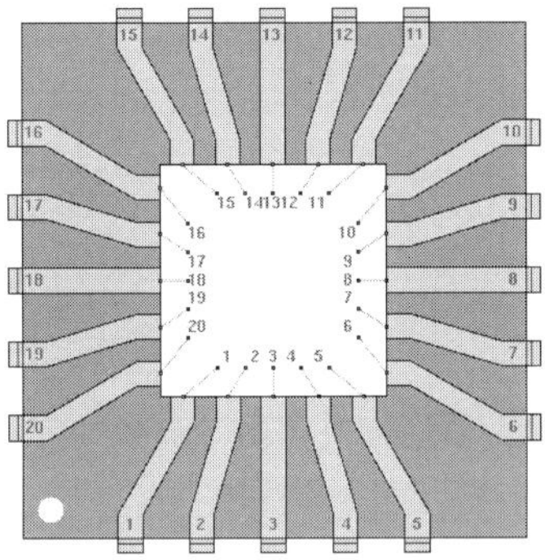

Figure 5.19. Physical structure of the package used in the simulations.

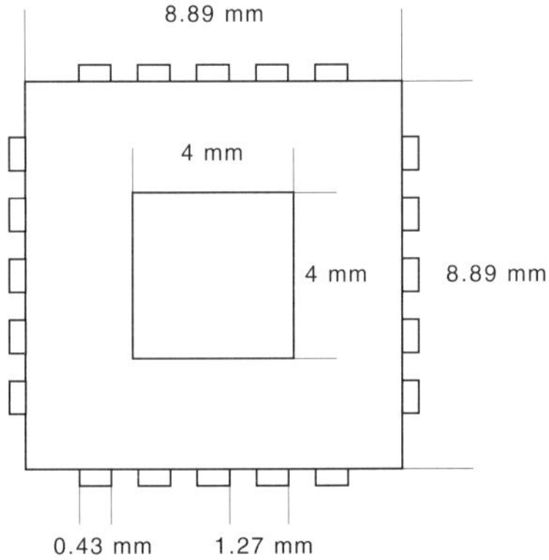

Figure 5.20. Dimensions of the PLCC package modeled.

In order to illustrate the effect on a typical package, a lumped RLC model of a PLCC with 20 leads has been generated using the tool RF IC Package Modeler from Cadence [7]. The dimensions and physical structure are shown in figures 5.19 and 5.20.

The frequency-dependent impedance can now be simulated by connecting an AC voltage source between, for example, bond-pins 1 and 2 and a terminating resistor emulating the power voltage source between the corresponding outer terminals[1]. The rest of the leads are connected to external ground. Figure 5.21 shows the impedance without decoupling capacitance, and with several decoupling capacitors. If no decoupling capacitance is present, the

[1]We are only concerned by the noise generated by the chip package. It is assumed that the voltage is perfectly regulated outside the package and this implies, among other measures, the use of decoupling capacitors at PCB level.

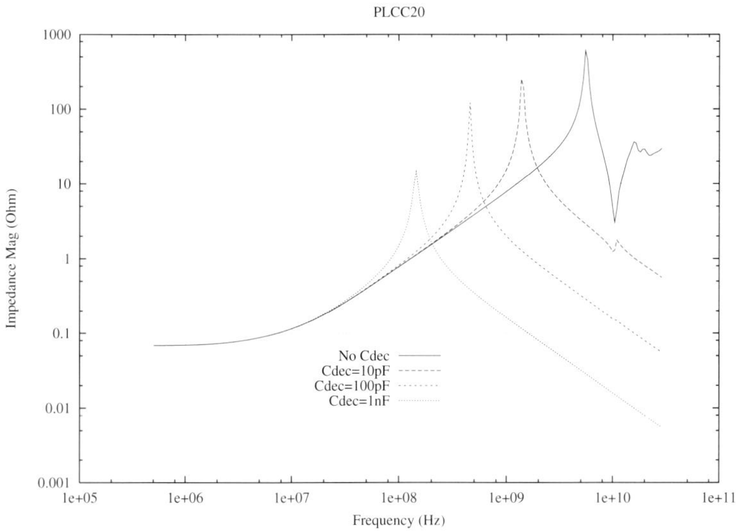

Figure 5.21. AC impedance without decoupling capacitance, and with several values of decoupling capacitance.

impedance increases because of the total effective lead inductance (except for some resonance peaks). The decoupling capacitance compensates for this increase by effectively shorting the inner terminals at high frequency, therefore reducing the impedance and, consequently, the noise generated. The bigger the capacitor is, the lower the impedance at high frequency.

Figure 5.22 shows the result in the time domain for simultaneous switching of nine CMOS inverters, with channel sizes $W_p = 7\,\mu\text{m}$, $W_n = 3\,\mu\text{m}$, $L = 0.35\,\mu\text{m}$. A significant reduction in noise amplitude is observed when there is a decoupling capacitance.

Note that there are also decoupling capacitors at PCB level, located as close as possible to the packages in the PCB. The rules for selecting the appropriate value and type of capacitor are similar because the problem is essentially the same. However, the capacitors at PCB level show relatively high parasitic series inductance, which cause its total impedance to increase at high frequency, therefore making the decoupling capacitor useless for such frequencies [8], [9].

In order to address the SSN problem at the package level, the decoupling capacitor must be integrated on- chip. Therefore, its associated series inductance is negligible, or, at most, comparable to the own on-chip power distribution network. The use of on-chip capacitors implies an increase in chip area and, correspondingly, in cost. However, it should be noted that an important amount of decoupling capacitance already exists as parasitic capacitance of PN junctions in P- and N-wells and in transistor drains. All the gates and blocks not

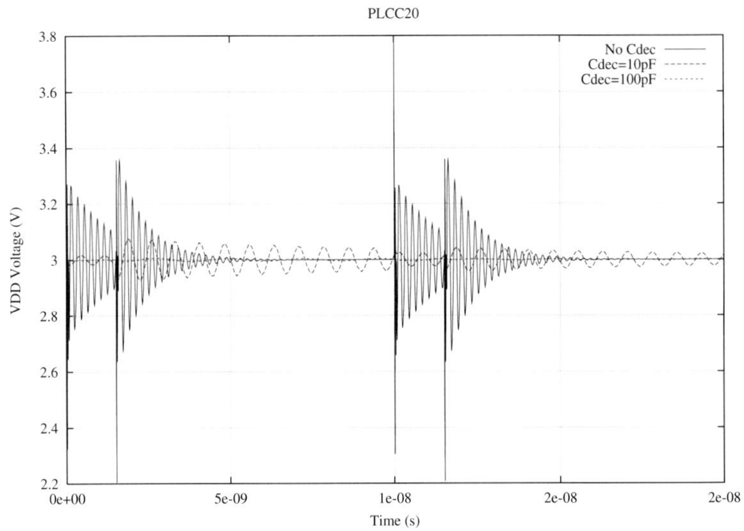

Figure 5.22. Transient simulation of the noise in the power supply, without decoupling capacitance, and with several values of decoupling capacitance. The noise is not distinguishable in this case for a capacitance of 100 pF.

switching contribute to this "free" decoupling capacitance. The problem with this capacitance is parasitic series resistance [10] (equivalent series resistance, or ESR) and not inductance. Figure 5.23 shows the difference in impedance for decoupling capacitors of 10 pF and 100 pF with an ESR of 10 Ω compared to the case of ideal capacitors. This resistance has as a benefit that the resonance is damped, but, at high frequencies beyond resonance, the behavior is worse (although still normally better than without any decoupling capacitance). Figure 5.24 shows the transient simulation with ESR and without ESR for the same 10 pF decoupling capacitor.

However, on-chip parasitic capacitance is not the solution to switching noise. The problem is that this parasitic capacitance can only function as decoupling capacitance for non-switching gates. As the number of switching gates varies with time, the amount of total capacitance is not fixed, and, what is worse, it is smaller when most needed, that is, when more gates are simultaneously switching. For this reason, it is necessary for high-performance circuits to integrate additional decoupling capacitance in spite of the cost in area. Decoupling capacitance can also be integrated into the package, next to the silicon die [11].

2.2.2 Switching of external nodes

However important, on-chip decoupling capacitance does not completely address all the noise generated in the chip. The reason is that the chip out-

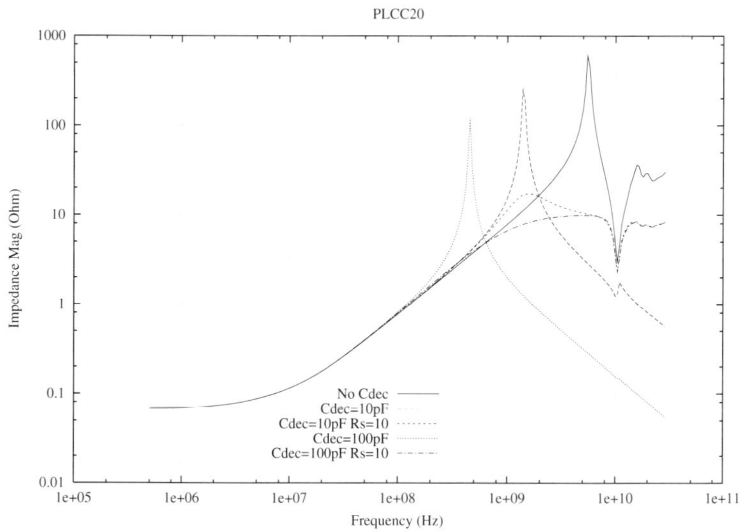

Figure 5.23. Effect of ESR in the value of AC impedance.

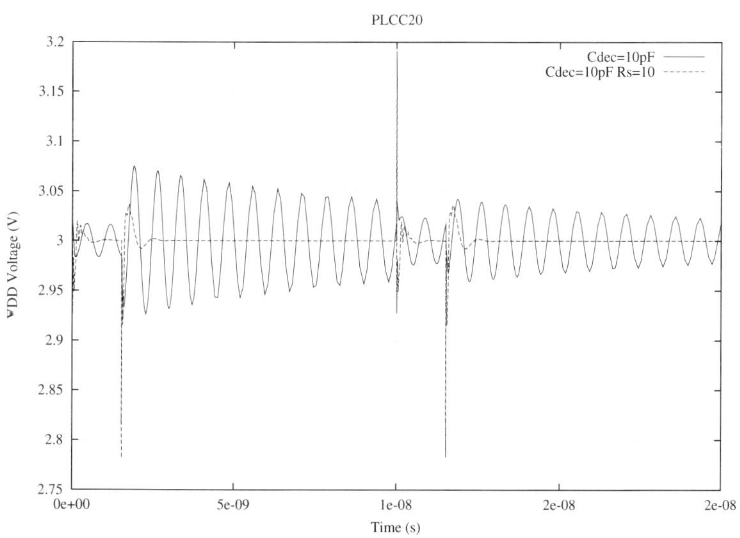

Figure 5.24. Effect of decoupling capacitance ESR in the power supply noise voltage transient waveform.

put buffers (those that are switching capacitances outside the package) are not affected by on-chip decoupling capacitance [12].

In the case of external switching nodes, all the current going out through the output lead to charge the external load must necessarily return through the external (package) power distribution network. Even though part of this current may have been supplied initially by on-chip capacitance, the current has to go through the whole package inductance and, therefore, no noise reduction can be observed.

Consequently, the case of output buffers is more critical from the point of view of noise. To reduce it, other strategies, like lead assignments and output buffer circuit design, must be tried. These are dealt with in the sections below.

2.3 Pin assignment

Pin assignment to physical package leads has a strong influence on the noise generated, for both internal and external switching [13]. There are two factors in pin assignment that influence noise. The first one is how many package leads are assigned to the power pin, P (connected to V_{DD} terminal of the power supply distribution) and to ground, G (connected to V_{CC} terminal). It is easy to see intuitively that the more leads dedicated to power and ground, the smaller the equivalent (*effective*) inductance of the respective connections, because the resulting effective inductance is the connection in parallel of several inductances. The mutual capacitance between leads also has a certain influence by determining the resonance frequencies.

The second, less obvious, factor with respect to pin assignment is how the P and G assigned leads are located with respect to each other. Remember that the loop inductance depends on the area of that loop. If the loop area is reduced, the inductance corresponding to that loop is also reduced. Considering again the case of internal switching, the current loop area is determined by the distance between one P lead and its closest G lead. As a consequence, if P and G leads are adjacent, the associated inductance is minimal and so is the generated noise [14].

As an illustration of the influence of these two factors, let us take the package model for a PLCC20 used in the previous section. Figure 5.25 shows the equivalent impedance of the package power distribution network for different lead assignments. Firstly, the case of a single P/G pair is shown for two different assignments: maximum distance between P/G leads (1 and 11), and adjacent P/G lead (1 and 2). Adjacent leads present a smaller impedance due to smaller equivalent or effective inductance, as denoted by the displaced curve at mid-frequencies in the log-log plot. The second class of curves correspond to four pairs of P/G leads assigned differently: in one case, all the P leads are located next to one another (1, 2, 3, 4) and at maximum distance from the G leads (11, 12, 13, 14). In the other cases, the P and G pairs are adjacent, but with contiguous pairs (1-2, 3-4, 5-6, 7-8), or separated, one pair in each side of the package (1,2, 6-7, 11-12, 16-17). As shown in figure 5.25, the DC value (dependent on equivalent resistance) is the same for all three cases, but as frequency rises (and effective inductance dominates the value of impedance), the

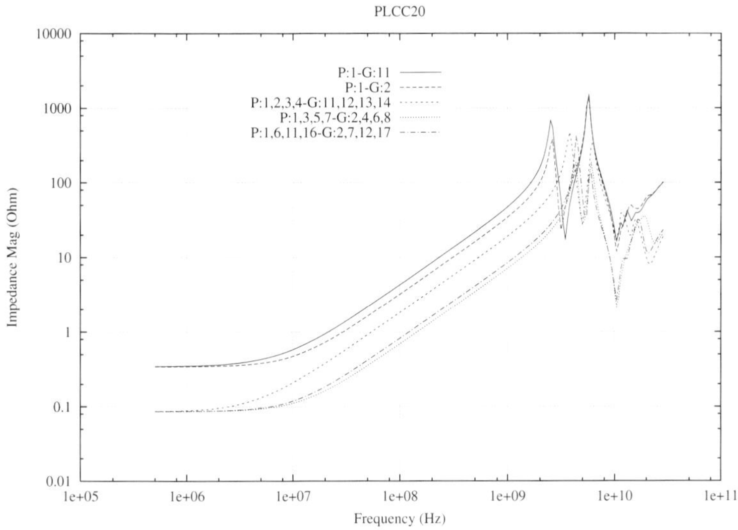

Figure 5.25. Influence of P and G assignment on the equivalent power distribution impedance.

cases with adjacent P/G pairs are better than the others. The best overall case in terms of equivalent impedance is the one with P and G lead pairs consecutively grouped. However, the case with all reference leads grouped together increases crosstalk coupling between signal package leads, and, therefore, an intermediate solution may be better.

In the case of logic blocks switching nodes outside the package, it is not possible to calculate the equivalent impedance because there are several possible paths. However, the same principle as in the case of internal node switching applies: increasing the number of package leads assigned to P and G reduces the effective inductance and, therefore, the noise will also decrease.

In order to illustrate this dependence, the following simulation examples are presented, using the same package model as in previous sections.

Figures 5.26 and 5.27 show the assignments used in the simulations. In both cases, leads number 1 to 10 are assigned to either quiet inputs (externally connected to ground, internally to the input of inverter blocks), or to P/G grouped in pairs. In one group of simulations (figure 5.26), the other leads (number 11 to 20) are externally connected to a switching voltage, and internally connected to the input of 9- inverter blocks, giving a total of 90 switching internal nodes. In the other group of simulations (figure 5.27) lead 11 is connected to a block of 9 inverters and the output of each inverter is connected to another inverter connected to leads 12 to 20, so, in this case, 9 outputs and one input block are simultaneously switching. For each group of simulations, the quiet input

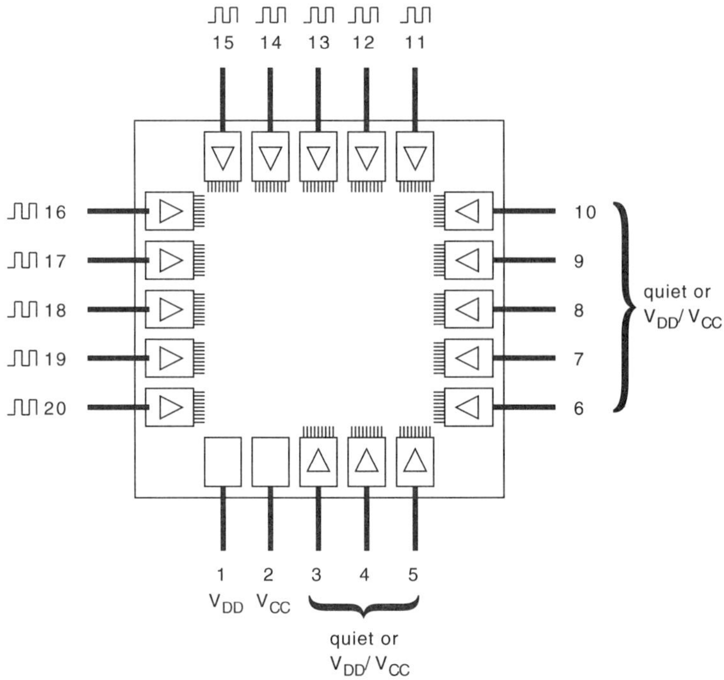

Figure 5.26. Pin Assignment example. Internal switching elements.

blocks are progressively replaced by connections to P or G, from one pair to 5 pairs, giving a total of 5 simulations.

Figure 5.28 shows the results of the simulations for the internal switching nodes. For each simulation, the maximum and minimum internal power voltage are given. Although it can be argued that it is an incomplete noise characterization, it gives an idea of how important switching noise is for each case. The same simulations have been repeated for different values of on-chip decoupling capacitance. The results show two main features: one, that the bigger the decoupling capacitance is, the smaller the noise generated. This illustrates the discussion about the internal impedance in function of decoupling capacitance for internal switching nodes given above. The second feature is that there is, in general, a reduction in noise when the P/G pairs increase in number. However, there are some irregularities because, when substituting one pair of quiet input blocks for a pair of P/G, there are two opposing effects: one is to reduce the effective inductance, and the other is to reduce the parasitic on-chip capacitance. Therefore, it is not easy to predict, a priori, whether or not the substitution will be beneficial from the point of view of noise.

Analyzing the results of the circuit with switching output nodes (figure 5.29) shows a clearer dependence of noise amplitude on the number of P/G pairs. As

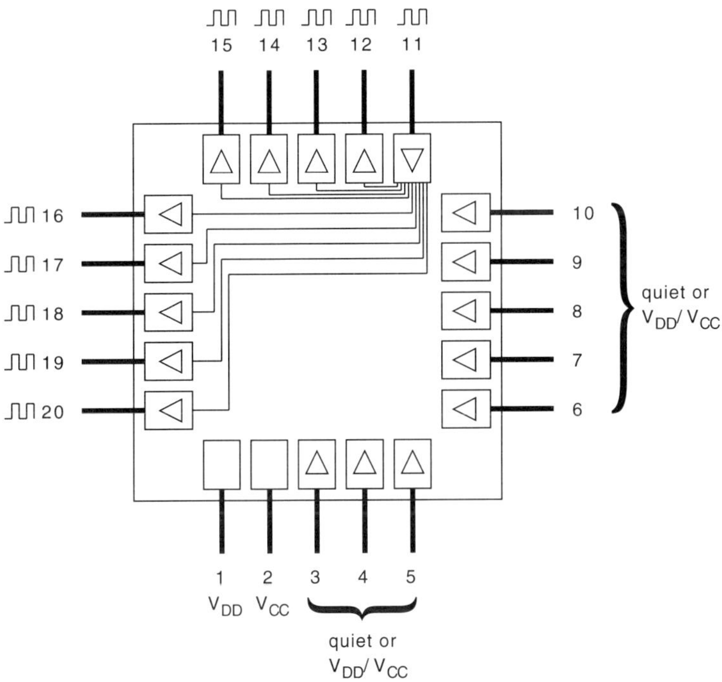

Figure 5.27. Pin Assignment example. External switching elements.

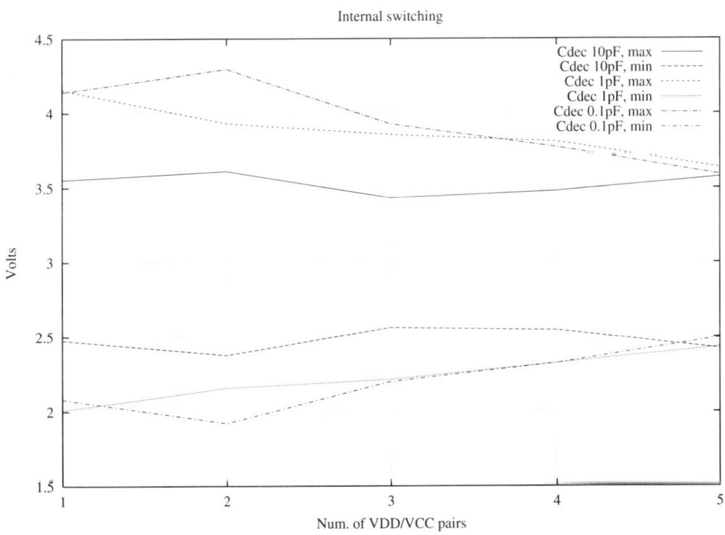

Figure 5.28. Maximum and minimum noisy power supply voltage for 90 simultaneously switching internal CMOS inverters.

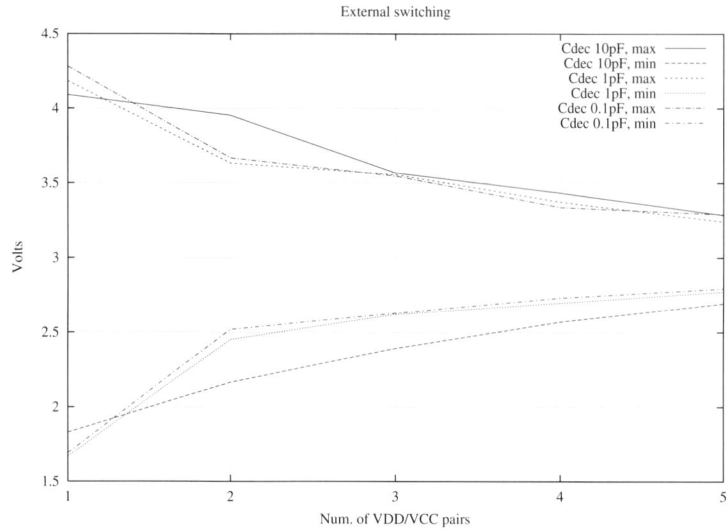

Figure 5.29. Maximum and minimum noisy power supply voltage for 9 simultaneously switching CMOS output inverters.

already discussed, the decoupling capacitance has no influence on the switching output noise, but here a certain dependence can be observed because there is a block of 9 switching inverters, which benefits from the decoupling capacitance.

2.4 Circuit techniques

In addition to improving the power distribution system, choosing an advanced package technology, adding decoupling capacitors and suitably assigning P and G pins to package leads, some improvements can be made at the circuit design level. There are several techniques for improving noise immunity or reducing the noise generated. Among these, we will briefly address four of them: use of differential signaling, controlling the skew between quasi-simultaneous transitions, specific output buffer design, and special digital architectures for low-noise generation.

2.4.1 Differential signaling

One of the circuit techniques is the use of differential signaling instead of a single wire interface [15]. The use of differential signals has two advantages. From the point of view of noise immunity, it has a clear advantage because any voltage reference fluctuation will equally affect both lines and is, therefore, a common mode noise that is cancelled by the differential receiver [8]. From the point of view of noise generation, differential drivers based on current steering also have an advantage over conventional CMOS static drivers, because the current is always constant, its time derivative is null and, therefore, they do not generate switching noise [16].

The increased noise immunity provided by differential signaling has two positive consequences: one is that the number of package leads dedicated to power and ground pins can be smaller. This advantage is compensated by the need to use two leads per signal I/O. The second consequence, more important, is that the voltage swing can be very small compared with single-ended signals that need to have a large S/N ratio.

One example of this type of signaling is the standard LVDS (Low Voltage Differential Signaling [17]). This low voltage swing implies that dynamic power dissipation is much smaller than conventional single end signaling and it is also faster. The reduced dynamic dissipation compensates the constant current consumption and it makes it a very attractive alternative for very high data rate transmission [18].

2.4.2 Controlled skewing of output drivers

As the generated noise increases with the number of simultaneously switching devices, one way of limiting the noise is to avoid simultaneity. This can be achieved by appropriately skewing the "simultaneous" transitions so that they are no longer simultaneous [19]. This solution obviously delays the bus switching because the bus value is not stable until all outputs have their definitive value. Additionally, it must be taken into account that switching noise shows an oscillatory behavior. Therefore, the skew ΔT between consecutive switching outputs must not be in phase with the noise because then the noise would increase instead of decrease. This certainly complicates the design of the skew control and makes this technique difficult to apply.

2.4.3 Output buffer design

Another field with possibilities for improvement is the design of chip output buffers. Remember that output buffers do not benefit from decoupling capacitors and the only way to decrease the noise they generate is by adding P/G assigned leads.

One technique already mentioned regarding the design of output buffers is the use of current-steering logic [20]. This logic style is differential by nature and therefore also shows high noise immunity.

There are other techniques that can be applied to conventional static single-ended logic [21], [19]. These techniques are based on minimizing the current derivative when a switching event occurs. First, let us analyze the switching characteristics to understand how this works.

Figure 5.30 shows a simplified schematic of the currents involved in the switching of an external node. The generated noise inside the package power

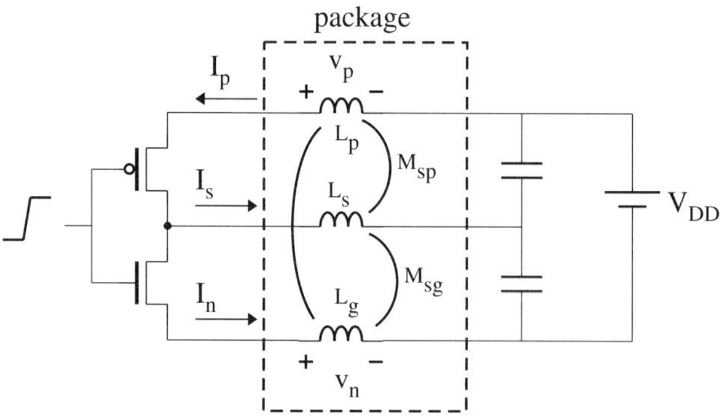

Figure 5.30. Noise generated due to one switching output buffer.

and ground terminals, represented as v_p and v_n, can be related to the current derivatives and approximately expressed as follows:

$$v_n = (L_g - M_{sg})\frac{dI_n}{dt} + (M_{sg} - M_{pg})\frac{dI_p}{dt} \quad (5.5)$$

$$v_p = (-L_p + M_{sp})\frac{dI_p}{dt} + (M_{pg} - M_{sp})\frac{dI_n}{dt} \quad (5.6)$$

Normally, in the switching process, there is a certain amount of through current (also known as overlapping or short-circuit current), due to simultaneous conduction of the NMOS and PMOS transistors during a certain interval. From expressions 5.5 and 5.6 it can be seen that if the overlapping current does not exist ($I_n = 0$ for a LH transition and $I_p = 0$ for a HL transition) the corresponding component derivatives ($\frac{I_n}{dt}$ or $\frac{I_p}{dt}$) will also be zero and the noise generated is therefore smaller. To design output buffers with no overlapping current, the usual technique is to use a structure based on a tri-state buffer, in which the PMOS and NMOS transistors are driven by the same signal, but delayed conveniently such that they are never ON at the same time (figure 5.31). The NAND gate has two series NMOS transistors and the NOR gate two parallel NMOS transistors, so the transition HL is much faster for the NOR gate driving the output NMOS than for the NAND gate. The complementary case happens for LH transitions. With this structure, the output buffer presents a small interval in a high-impedance state, but the overlapping current is nearly zero and the resulting noise generated is reduced with respect to the case of a conventional output buffer. The penalty is an increase in complexity and, most importantly, an added output delay.

Also, it should be noted that, for a conventional output buffer, the overlapping current is also reduced if the output is much slower than the input. As a general rule then, slower outputs will generate a smaller amount of noise.

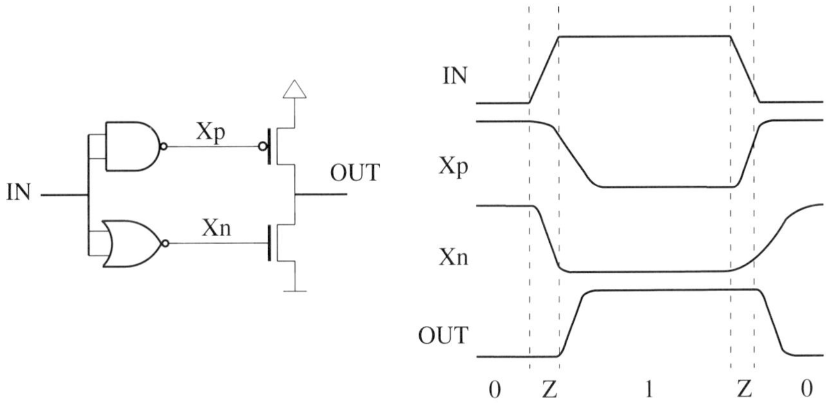

Figure 5.31. Signal operation for the modified low SSN output buffer.

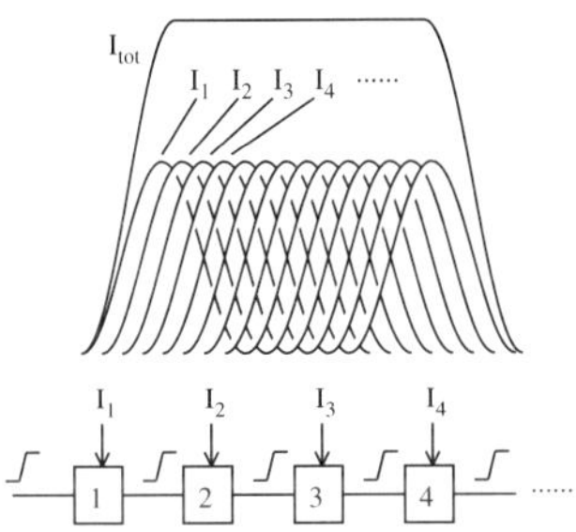

Figure 5.32. Total current consumption giving a nearly zero current derivative during circuit operation.

2.4.4 Special architectures

Finally, another way to design low SSN noise circuits is by trying to control the current waveform of the switching blocks [22]. Taking into account that the noise is in direct relation to the current derivative, the way to reduce noise is to reduce this current derivative. Current steering logics achieve this goal at the price of constant current consumption. An intermediate solution is to design in such a way that the current is nearly constant during the time of operation, by assuring that the overlapping of the blocks' current consumption is constant. To achieve this, the delay of each block must be the same, as well as the current peak of each individual block (figure 5.32). Therefore, the current derivative is non-zero only when the operation starts and ends, instead of having several peaks for the switching of the different blocks. This technique is well suited to specific regular structures, like multipliers [23], but may be more difficult to apply to general structures.

References

[1] Vittal A. and Marek-Sadowska M. "Crosstalk reduction for VLSI". *IEEE Trans. on Computer Aided Design of Integrated Circuits and Systems*, 16(3), March 1997.

[2] ITRS. International Roadmap for Semiconductors: 2001, 2001.

[3] Avant! Corp. *Star-Hspice 2001.2*, June 2001.

[4] Alpert C. J., Devgan A., and Quay S. T. "Buffer insertion for noise and delay optimization". *IEEE Trans. on Computer Aided Design of Integrated Circuits*, 18(11), November 1999.

[5] Rubio A., Pons J., and Anglada R. "A Crosstalk Tolerant Latch Circuit". *IEE Proceedings on Circuits, Devices and Systems*, 38(1), February 1992.

[6] Harada A., Kaizu K., Yamanaka N., and Kawamura T. "Development of plastic chip scale package for ATM switching systems". In *IEEE/CPMT Electronics Packaging Technology Conference*, 1998.

[7] Cadence Inc. RF IC Package Modeler, 2003.

[8] Johnson H.W. and Graham M. *High-speed digital design: a handbook of black magic*. Prentice Hall, 1993.

[9] Smith L.D., Anderson R.E., Forehand D.W., Pelc T.J., and Roy T. "Power distribution system design methodology and capacitor selection for modern CMOS technology". *IEEE Trans. on Advanced Packaging*, 22(3), August 1999.

[10] Larsson P. "Parasitic resistance in an MOS transistor used as on-chip decoupling capacitance". *IEEE J. of Solid State Circuits*, 32(4), April 1997.

[11] Val C. M. and Martin J. E. "A New Chip Carrier for High Performance Applications: Integrated Decoupling Capacitor Chip Carrier (IDCC)". *IEEE Trans. on Components, Hybrids and Manufacturing Technology*, 6(3), September 1983.

[12] Bakoglu H. B. *Circuits, Interconnections and Packaging for VLSI*. Addison-Wesley, 1990.

[13] Lin L. and Prince J. "SSO noise electrical performance limitations for PQFP packages". *IEEE Trans. on Components, Packaging and Manufacturing Technology–Part B*, 20(3), August 1997.

[14] Rainal A. J. "Computing Inductive Noise of Chip Packages". *AT&T Bell Laboratories Technical Journal*, 63(1), January 1984.

[15] Rainal A. J. "Eliminating Inductive Noise of External Chip Interconnections". *IEEE Journal of Solid State Circuits*, 29(2), February 1994.

[16] Allstot D.J., Kiaei S., and Zele R.H. "Analog logic techniques steer around the noise". *IEEE Circuits & Devices Magazine*, 9(5), September 1993.

[17] TIA/EIA-644A. Electrical characteristics of low-voltage differential-signalling (LVDS) interface circuits, 1996.

[18] Boni A., Pierazzi A., and Vecchi D. "LVDS I/O interface for Gb/s-per-pin operation in 0.35μm CMOS". *IEEE J. of Solid State Circuits*, 36(4), April 2001.

[19] Senthinathan R. and Prince J. *Simultaneous switching noise of CMOS devices and systems.* Kuwer Academic Publishers, 1994.

[20] Ng H.-T. and Allstot D.J. "CMOS current steering logic for low-voltage mixed-signal integrated circuits". *IEEE Trans. on Very Large Scale Integration Systems*, 5(9), September 1997.

[21] Leung K. "Controlled slew rate output buffer". In *IEEE Custom Integrated Circuits Conference*, 1988.

[22] Aragones X., Gonzalez J.L., and Rubio A. *Analysis and solutions for switching noise coupling in mixed-signal ICs.* Kluwer Academic Publishers, 1999.

[23] Gonzalez J.L. and Rubio A. "Low delta-I noise CMOS circuits based on differential logic and current limiters". *IEEE Trans. on Circuits and Systems I: Fundamental Theory and Applications*, 46(7), July 1999.

Chapter 6

NOISE DETECTION AND TESTING IN LOGIC CIRCUITS

The effects of crosstalk and SSN on circuit behavior are very similar: either they cause an unwanted pulse (spurious signal) in a quiet line, or a change in transition delay in a switching line. However, in the case of crosstalk, unlike SSN, the generation of these effects is localized. The magnitude of these perturbations depends on the values of the electrical parameters involved: lines, drivers and load capacitances. A great deal of work has been published to evaluate the amplitude and width of the spurious signal produced by crosstalk, especially in the case of capacitive crosstalk and, more recently, including inductive crosstalk [1], [2].

Once the actual circuit is built and manufactured, it is necessary to check whether the existence of crosstalk and SSN can alter its functionality. Therefore, just as with any physical defect, the circuits must be tested for noise. As with traditional tests, the problem consists of applying a set of inputs that maximize the possible error. Here is where crosstalk and SSN tests differ, because the excitations for maximum crosstalk are not generally the ones that produce maximum SSN.

This chapter presents the work on noise testing, with a special emphasis on crosstalk tests. Where appropriate, the relationship with SSN tests will also be pointed out.

1. The noise detection problem

After the circuit is manufactured, it must be tested and potential crosstalk problems must be analyzed from an experimental point of view. Due to the enormous reduction in dimensions in today's technologies, crosstalk and SSN is becoming more important. The question currently arising is: if the layout has been analyzed, why is the detection of crosstalk necessary after manufacturing the integrated circuit? Why must post-manufacture detection be carried out? Various facts justify a positive answer. While it is clear that all efforts must be made at the design stage to prevent noise-related problems, the great complexity of modern chips involves the analysis of hundreds of thousands of interconnections. This analysis can only be carried out with simplified models, on a reduced number of interconnection groups which are potential candidates for significant crosstalk effects. The design tools currently handling this issue, guiding the routing phase of the design, are based on RC models, neglecting inductance. As discussed above, inductance can be important in certain cases and, therefore, the crosstalk effect in these cases may be underestimated. Besides, capacitance values are extracted or estimated based on simplified formulas. And, even when these necessary simplifications are close to reality, process parameter fluctuations may induce an increase in the effect previously calculated, which will only appear in the field. Therefore, efficient testing methods must be devised which consider noise as a fault that must be detected experimentally. An indication of the interest and importance of the problem of detecting noise at post-manufacturing stages is the inclusion of specific sessions on crosstalk detection and testing in today's most important test-related technical and scientific conferences. In this respect, from the point of view of circuit testing, noise may be considered as a soft fault, in the same family as single events caused by ionizing radiation effects.

The main tool in the detection of noise after fabrication is ATPG (automatic test pattern generation), which consists of applying certain algorithms for finding the appropriate vectors that may produce noise and allow its propagation to an external output or to an internal memory cell, where it can be observed. In the following section, some basic ideas and algorithms for generating test patterns in general will be commented on. After that, specific algorithms for the noise detection problem will be presented.

2. Brief introduction to the testing of logic circuits

The testing of integrated circuits is considered to be of strategic importance. The great complexity in the operation of modern integrated circuits makes the alternative of checking its functionality by exhaustively applying all the possible input combinations impossible. Therefore, the structural test approach is taken, consisting of trying to find structural defects, that is, those located in some specific part of the circuit. In this way, the required number of inputs is very much reduced compared with a functional test approach. For a structural test, the following aspects must be considered:

Defects, faults and fault models In order to clarify the objectives of the testing process, it is necessary to define two different concepts: defect and fault. A defect refers to a change in physical structure from the designed specifications. For example, a short-circuit between two different lines due to a spot on the mask, or a short in the gate oxide of an MOS transistor. A fault refers to a type of misbehavior of the circuit, which may correspond to several different defects. Test strategies are therefore oriented towards faults, which may be modeled at different levels of abstraction, from logical to physical. The fault model needs to be accurate to describe the response of the circuit, but simple enough to devise a rapid testing strategy. Examples of fault models include the stuck-at fault (logical level), or the bridge fault (physical/electrical level).

Observable magnitudes Each fault model induces a change on a set of given variables, and these variables (called observables) are the ones monitored to detect a problem in the circuit. The test technique is related to the type of observable considered and it consists of all the necessary procedures for checking and monitoring this magnitude. Examples of different observable magnitudes are output logic values, temperature, current consumption, input/output delay, etc.

Test vectors Given a certain fault, the problem of testing is to detect it by monitoring the observable magnitude. There will be certain circuit excitations for which no abnormal behavior is observed, and it is therefore necessary to find precisely those excitations that produce abnormal behavior for a particular fault at a specific location. The circuit inputs that produce such excitation are what are called test vectors. The procedure for finding test vectors (test pattern generation) is carried out before applying the test, by using several algorithms.

2.1 ATPG for stuck-at faults

The techniques presented in a later section for addressing the detection and testing of crosstalk faults are based on the stuck-at fault model. The stuck-at fault is a logic-level fault model that consists of assigning a fixed logic value at a certain node of the circuit, independently of the applied inputs. If the fixed value is a logic 1, it is a stuck-at-1 fault and it is a stuck-at-0 fault if the fixed value is a 0. This fault model is somewhat unrealistic, but nevertheless it provides important coverage of real defects, and it is still widely used, in conjunction with more realistic fault models.

The test pattern generation procedure for stuck-at faults can be split in two phases:

Controllability It consists of finding an input vector that puts the node where the stuck-at fault is at a logic value opposite to the stuck-at fault considered.

Observability After the controllability phase, the discrepancy in the faulty node must be propagated to an external output (primary output) so it can be observed.

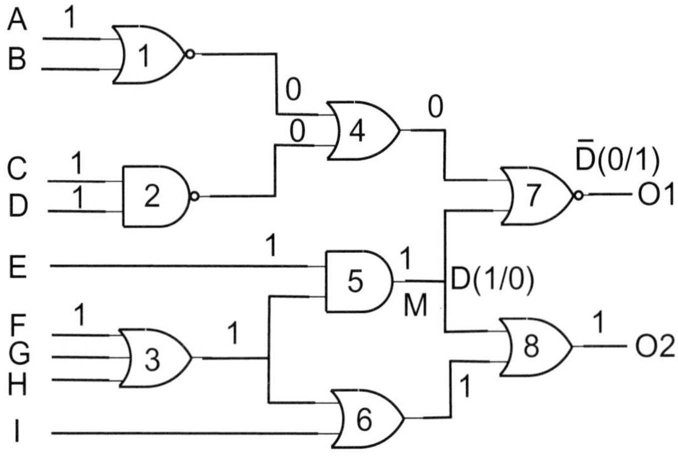

Figure 6.1. Example of stuck-at fault detection.

Figure 6.1 shows an example of a how a stuck-at fault is detected. In the logic circuit in the figure, a stuck-at-0 fault at node M has to be detected. We want to find an input vector such that a discrepancy from the normal output due to this fault is observed.

First, the controllability phase consists of putting node M at logic 1. This can be achieved, for example, putting a 1 in primary (external) inputs E and F. The rest of the primary inputs are unassigned for the moment. Now, the observability phase has to achieve the propagation of the value at node M to one of the primary outputs, $O1$ or $O2$. To indicate that there is a discrepancy due to a fault, the value at node M is denoted as D. The propagation of D to output $O2$ is blocked because the controllability phase fixed the output of gate 6 to 1, and therefore, $O2$ is also fixed to 1. Therefore, the only output where D can be observed is $O1$. In order to allow this propagation, one possibility is putting $A = C = D = 1$, so that the opposite value of node M is observed: a logic 0 if there is no fault and a logic 1 if there is a fault. This is denoted as logic value $\bar{D}$. Note that a 5-valued logic is used to describe the circuit in the presence of faults, the five possible values being $\{1, 0, D, \bar{D}, X\}$.

In general, for complex circuits it is almost impossible to find the test vectors manually as in the previous example. Several algorithms were developed to generate the test patterns. The classical algorithms are the D-algorithm [3], the PODEM algorithm [4] and the FAN algorithm [5]. As the algorithm explained later for the crosstalk problem is based on PODEM, we concentrate on this algorithm to explain some basic concepts.

2.2 Backtrace and backtrack procedures in PODEM

There are two main procedures in the PODEM algorithm: backtrace and backtrack. In general, both controllability and observability phases have a given objective, consisting of putting a desired logic value in the corresponding node. The backtrace procedure traces the objective node back to one of the primary inputs and assigns a logic value to it so that the objective is achieved, in function of the type of logic gates making up the path. There are several possible paths from the objective node to the input, and the selection of one particular path is made depending on a set of criteria known as heuristics. The heuristics can be varied: random, deterministic, probabilistic, etc. Generally, more than one backtrace process is needed to achieve the desired objective. When one backtrace process is completed (a determined value is assigned to a primary input), three situations can arise:

1 The objective node has the desired logic value. Therefore, the objective is achieved, and a new objective is chosen.

2 The objective node is still unassigned. In this case, a new backtrace operation is needed, with the same objective.

3 The objective node is set to a wrong value. This situation is possible because, when setting a primary input, several paths are activated that may arrive at the same node. In this case, a backtrack operation is needed, as explained below.

A backtrack operation is needed in the case of conflict between the objective and the result obtained. The way to solve the conflict is, firstly, to set the last primary input assignment to the opposite value. If the conflict situation still persists, this last assignment is unset and the next last assignment is set to the opposite value. This operation is repeated until there is no conflict, that is:

- The objective node is unset, and therefore, a new backtrace procedure can be applied, or,

- The objective node is set to the desired value, and therefore a new objective can be chosen.

Figure 6.2 illustrates the backtrace and backtrack procedures on a simple circuit. In this case, the objective is to set node I to 1. The backtrace procedure starts and, assuming the heuristic, selects the path given by nodes $C-F-H-I$. It can be seen that the value 1 at input C is the least intrusive and, therefore, this is the value assigned by this first backtrace procedure. At this stage, all other nodes including the objective node I are left unassigned (value X). A second backtrace procedure selects path $A-D-F-H-I$, and assigns a 0

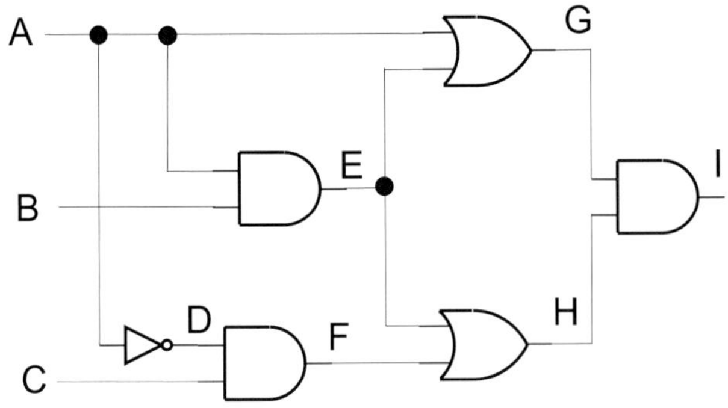

Figure 6.2. Circuit example to illustrate backtrace and backtrack procedures.

to A. With this new assignment, $D = 1$, $F = 1$, $H = 1$, $E = 0$, $G = 0$ and $I = 0$. Therefore, a conflict situation arises. A backtrack operation is called, flipping the value of A (the last assigned input) and setting it to 1. This new assignment sets the objective node I to X. Now, a new backtrace operation selects path $B - E - H - I$ (the only one left) and sets B to 1. This last assignment achieves the desired objective of putting I to 1.

Once the controllability phase has been carried out, observability consists of allowing the propagation of the discrepancy (value D or $\bar{D}$) from the node where the stuck-at fault is located to one of the circuit's primary outputs, where it can be observed. This is achieved by making the subsequent gates transparent: setting logic 1 at the inputs (other than the one that we want to propagate) of AND and NAND gates, and a logic 0 to the inputs of OR and NOR gates. Therefore, there is a succession of objectives until the observability phase is completed, each with its backtrace and backtrack operations. The selection of an efficient heuristic is very important in order to obtain the test vector more quickly.

The success of a test pattern generation process is measured in terms of fault coverage. Fault coverage is defined as the percentage of faults detected of the total faults considered. Some of the non-detected nodes are due to inefficiency of the algorithm and some faults can be truly undetectable because of the circuit topology.

2.3 SSN detection problem

As already mentioned, crosstalk and SSN can be thought of as a fault, using test terminology. In this case, the fault is not necessarily due to a defect.

In the case of SSN, the main difference with respect to a classical fault model is that the fault is not located in a particular node, and thus, the controllability phase is different. The problem is not trying to create a discrepancy in a certain node, but to excite the circuit so that maximum SSN is produced and, after that, propagate the effect from the node considered for the fault to a primary output (observability phase). The list of faults to be considered can be exhaustive, as the whole circuit is approximately equally affected by SSN. There are two faults to be considered: spurious signals and delay faults [6].

2.4 Crosstalk detection problem

In the case of crosstalk, of all the nodes in the circuit, only groups of nodes that are close to one another need to be considered for the detection problem. Therefore, in order to create a fault list for which to generate test vectors, a layout observation is required. This analysis can be carried out using existing tools that automatically select "hot spots" depending on the ratio between coupling and ground capacitance, and driver resistance ratio, for example. These tools are already used in the layout design stage to assist with the placement and routing process.

Once the fault list has generated, the problem is similar to a classical stuck-at test: for a victim line the controllability phase consists in causing a transition in its neighbor line and the observability phase is almost the same as in classical test. An algorithm for the detection of spurious signal faults [7], [8] is presented in the following section.

3. Crosstalk-induced spurious signal detection

The fault model considered for the crosstalk problem is a spurious signal. The delay fault associated with crosstalk will be discussed later on. The test strategy is based on a logic level and, therefore, the spurious signal produced is assumed to have enough amplitude to be interpreted by logic gates as a small pulse with amplitude equal to V_{DD} and width T. The observable magnitude and test technique are the same as in classical testing of stuck-at faults: the observation of their logical values at the primary outputs.

The controllability phase consists of causing the crosstalk with a transition (therefore, two input vectors are needed) and the observability phase is the same as in a stuck-at fault test. Note that the second vector after the transition is the one that must enable the propagation of the fault to some of the outputs.

A logic characterization of the crosstalk fault is needed so the following logic variables are defined:

- logic 1.
- logic 0.
- X. Not assigned.
- *TU*. Up transition (0 to 1) at time t.
- *TD*. Down transition (1 to 0) at time t.

AND	0	1	TU	TD	P0	P1	TUD	TDD	X
0	*0*	*0*	*0*	*0*	*0*	*0*	*0*	*0*	*0*
1	*0*	*1*	*TU*	*TD*	*P0*	*P1*	*TUD*	*TDD*	*X*
TU	*0*	*TU*	*TU*	*0*	*TUD*	*P1*	*TUD*	*P1*	*X*
TD	*0*	*TD*	*0*	*TD*	*TD*	*0*	*0*	*TD*	*X*
P0	*0*	*P0*	*TUD*	*TD*	*P0*	*0*	*TUD*	*TD*	*X*
P1	*0*	*P1*	*P1*	*0*	*0*	*P1*	*0*	*0*	*X*
TUD	*0*	*TUD*	*TUD*	*0*	*TUD*	*0*	*TUD*	*0*	*X*
TDD	*0*	*TDD*	*P1*	*TD*	*TD*	*0*	*0*	*TDD*	*X*
X	*0*	*X*	*X*	*X*	*X*	*X*	*X*	*X*	*X*

Table 6.1. AND function with 9-valued algebra.

OR	0	1	TU	TD	P0	P1	TUD	TDD	X
0	*0*	*1*	*TU*	*TD*	*P0*	*P1*	*TUD*	*TDD*	*X*
1	*1*	*1*	*1*	*1*	*1*	*1*	*1*	*1*	*1*
TU	*TU*	*1*	*TU*	*1*	*1*	*TU*	*TU*	*1*	*X*
TD	*TD*	*1*	*1*	*TD*	*P0*	*TDD*	*P0*	*TDD*	*X*
P0	*P0*	*1*	*1*	*P0*	*P0*	*1*	*P0*	*1*	*X*
P1	*P1*	*1*	*TU*	*TDD*	*1*	*P1*	*TU*	*TDD*	*X*
TUD	*TUD*	*1*	*TU*	*P0*	*P0*	*TU*	*TUD*	*1*	*X*
TDD	*TDD*	*1*	*1*	*TDD*	*1*	*TDD*	*1*	*TDD*	*X*
X	*X*	*1*	*X*	*X*	*X*	*X*	*X*	*X*	*X*

Table 6.2. OR function with 9-valued algebra.

NOT	0	1	TU	TD	P0	P1	TUD	TDD	X
	1	*0*	*TD*	*TU*	*P1*	*P0*	*TDD*	*TUD*	*X*

Table 6.3. NOT function with 9-valued algebra.

- *P1*. Positive pulse (0-1-0).
- *P0*. Negative pulse (1-0-1).
- *TUD*. Delayed up-transition, equal to $TU(t+T)$.
- *TDD*. Delayed down-transition, equal to $TD(t+T)$.

Although values *TUD* and *TDD* may not seem necessary, they are needed to complete the truth tables of the different logic functions. For example, *TUD* is obtained with the AND of *P0* and *TU*. As an illustration, tables 6.1, 6.2 and 6.3 show the 9-value truth tables of AND, OR and NOT functions.

Using this 9-value algebra, a modification of the PODEM algorithm allows the generation of test vectors for the detection of crosstalk spurious signals.

3.1 Crosstalk fault controllability and observability

From a post-layout examination, a list of crosstalk faults is obtained, using what is called inductive fault analysis (IFA). The result is a list of pairs of nodes, (X,Y), where the first node is the affecting node and the second node is the affected node (where the fault actually is). Note that in some cases, there may be a possible fault in both coupled lines, so the same pair of nodes may appear twice in the list in a different order. On the other hand, some pairs of nodes may only appear once, indicating that one node of the pair is susceptible to crosstalk, while the other is not. This is possible, for example, in the case that the respective drivers are very different, as explained in chapter 3.

Considering a crosstalk fault at node Y due to a transition at node X, denoted here as $C(X, Y)$, there are two possibilities for detecting it at some output:

- Causing a 0-to-1 transition at node X and propagating the resulting value in Y to some output. In this case, a first input vector has to set both X and Y to 0 (note that if Y was at 1, the spurious signal would be above V_{DD} and subsequent logic gates would interpreted as logic 1, not as a spurious signal). For this first vector, only the controllability phase applies. After that, a second input vector has to keep Y to 0, and set X to 1. For the second input vector, both phases (controllability and observability) have to be carried out, first to set the correct values and then to propagate the resulting fault (of value *P1* at node Y) to some of the outputs.

- Causing a 1-to-0 transition at node X. The initial values for the first input vector at nodes X and Y have now to be 1 (controllability phase only). The second vector needs to set X to 0 and keep Y to 1, and then propagate the fault (controllability and observability).

Therefore, the problem of detecting spurious signals can be solved with an extension of the classical stuck-at fault test, for which there are well-known algorithms. Before explaining the algorithm used and the results obtained, we show the procedure sketched above with an example. Figure 6.3 shows a circuit where M is the affecting line and N the affected line (therefore, a fault $C(M, N)$). We consider the detection by forcing a 1-to-0 transition at node M. Therefore, the objective for the first vector is $M = 1$, $N = 1$. This objective can be reached with primary inputs set to: $A = E = F = X$, $B = C = 1$, $D = 0$. This is the first vector. The second vector has an initial objective of $M = 0$, $N = 1$. By now setting $C = 0$, this first objective is met, so the values of nodes M and N in the 9-valued algebra are $M = TD$ and $M = P0$. This

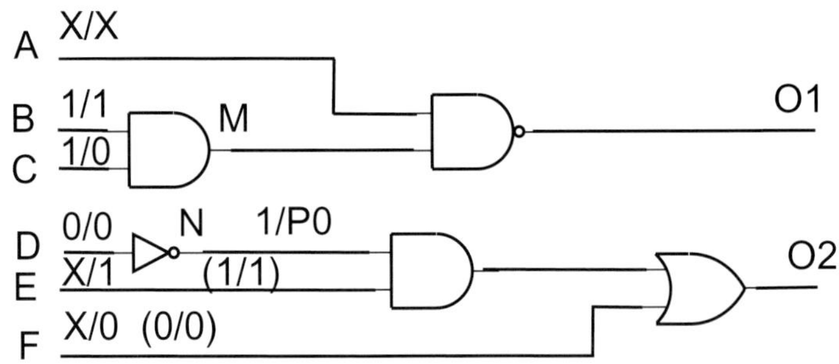

Figure 6.3. Circuit example to illustrate crosstalk test pattern generation.

P0 must be propagated, that is, either a *P0* or a *P1* value must be observable at one of the outputs at the end of the observability phase for the second input vector. To start the observability phase, the first objective is to set $E = 1$ and then, $F = 0$ (as in this example, these are primary inputs, this is achieved immediately). These two primary inputs were unset in the first vector and, in order to avoid unwanted transitions, the first vector's values for E and F are forced to the same values as in the second vector. As A is also unset, its value is also forced to be the same in both vectors, 0 or 1. Finally, there are two possible pairs of input vectors $(ABCDEF)$ that can be applied to observe a *P0* at the output: $(011010) \to (010010)$ or $(111010) \to (110010)$.

3.2 ATPG for crosstalk

An ATPG algorithm was implemented, enabling the extraction of some results concerning the testability of spurious signals [8]. Other algorithms and techniques are presented in a later section.

The algorithm begins with a list of crosstalk faults, as already mentioned in the example. In addition, it is assumed that each crosstalk fault has a maximum number of gates it can traverse (the penetration depth concept presented in chapter 3). Therefore, each node has an associated number, *MNG*.

The algorithm is divided in three different phases:

Phase 1 This first phase obtains the first of the pair of vectors, setting the value of the fault's pair of nodes to the same value. The initial objective (both nodes to 1 or both nodes to 0) is determined by a SCOAP-type heuristic [9]. If this initial objective cannot be reached, the objective is changed. If the new objective again cannot be reached, the fault is classified as undetectable.

Phase 2 This phase consists of finding the second vector setting the victim node at the same value as in the previous phase and the affecting node at an opposite value, thus causing a transition in this node.

Phase 3 The third phase sets unused primary inputs of the second vector in order to propagate the produced spurious signal. This problem of propagation is equivalent to the propagation of a D (or $\bar{D}$) value of the stuck-at model. Now *P0* or *P1* is propagated, using a similar algebra to the one for the stuck-at problem. In this phase, the limit on the propagation of the signal is taken into account by increasing the value of a counter variable each time a gate is traversed in the path to the output and decreasing it when the path is unable to propagate the signal (backtrack process). When this variable reaches a certain limit (*MNG*, representing the SS penetration depth), then a new path is chosen.

Figures 6.4, 6.5 and 6.6 show the charts for each of the phases. It can be observed that Phases 1 and 2 follow the same structure as both of them correspond to controllability problems. In Phase 3, the use of a counter for the number of gates ensure that the selected path to the output contains a lower number of gates than the spurious signal penetration depth *MNG*.

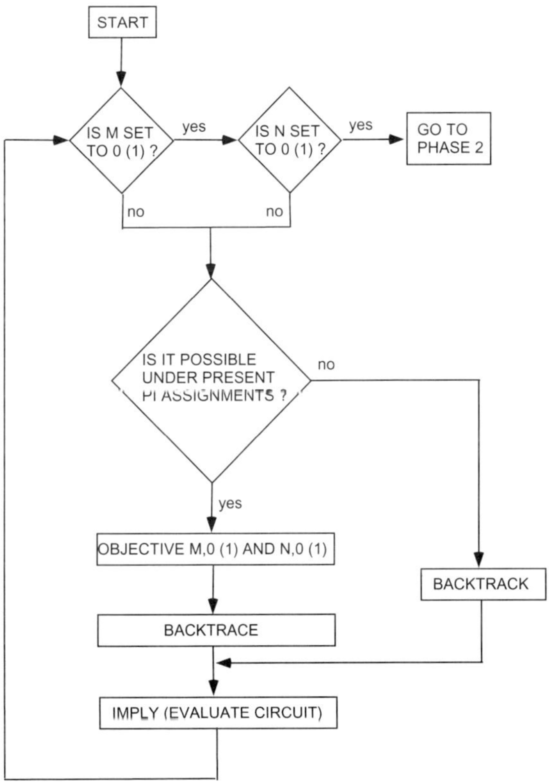

Figure 6.4. Phase 1 in ATPG process.

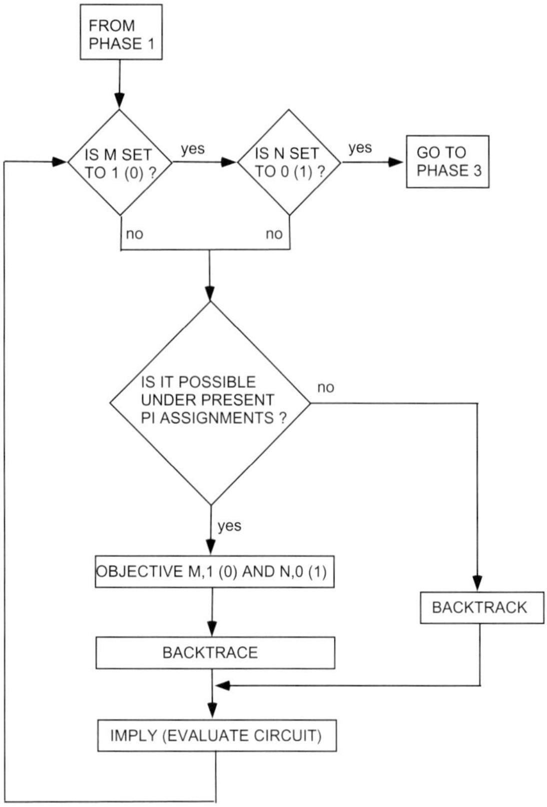

Figure 6.5. Phase 2 in ATPG process.

3.3 Results

Some results of applying the above algorithm are presented here. The circuits used to evaluate detectability are all combinational circuits, used as test benchmark circuits, known as ISCAS85 circuits [10]. We worked from a gate-level description without a layout, so to test the algorithm, 200 random faults were considered. The fault coverage (percentage of detected faults) is obtained from the algorithm. In order to see the influence of penetration depth on the detectability of crosstalk faults, different values of *MNG* are considered: 2, 5, 10 and without limit. The results are shown in table 6.4.

It can be noted from the results table that the influence of *MNG* is quite significant: for a small *MNG* the fault coverage drops dramatically for some circuits compared to the case of unlimited propagation. In the case of crosstalk faults, this is good news, because it means spurious signals with small penetration depth will have a small probability of reaching an output and affecting other circuits. On the other hand, a *MNG* of 10 gives almost the same results as

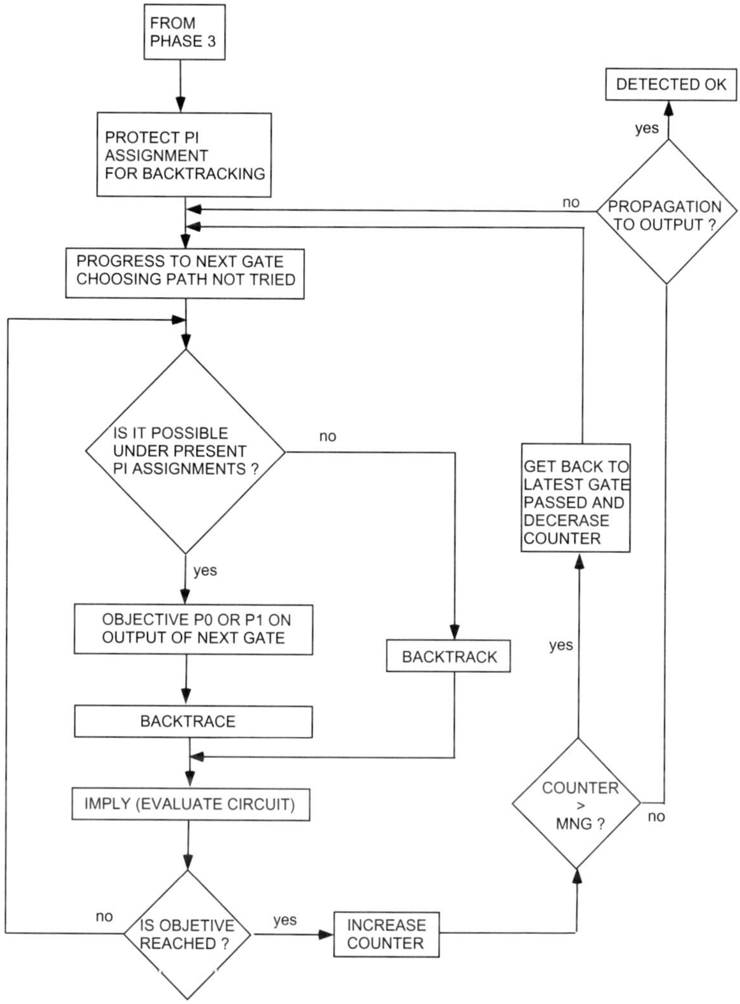

Figure 6.6. Phase 3 in ATPG process.

unlimited propagation for the ISCAS circuits. In summary, taking into account the penetration depth, the number of actual faults to be considered may be very much reduced.

The results presented are for combinational circuits. For sequential circuits, the circuit can be partitioned in a combinational part driving a sequential part, so that the propagation of *P0* or *P1* is to the input of the sequential block. After that, Phase 3 would be re-applied in order to propagate a possible erroneous state to a primary output.

CIRCUIT	*MNG=2*	*MNG=5*	*MNG=10*	*UNLIMITED*
C432	40.5	78.0	98.0	98.0
C499	61.5	84.5	92.5	92.5
C880	56.0	99.5	100	100
C1355	81.5	94.5	97.5	97.5
C1908	36.0	80.0	99.0	99.0
C2670	56.5	96.5	99.0	99.0
C3540	62.0	78.0	82.0	84.5
C6288	78.5	94.5	94.5	94.5

Table 6.4. ATPG fault coverage percentage for ISCAS combinational circuits, considering 200 random faults.

4. Other test techniques

The previous section illustrated an alternative to test circuits for spurious signals induced by crosstalk. This topic, which is of practical importance because of the increasing effect of interconnects in deep-submicron circuits, is the subject of considerable activity from several research groups worldwide. This section shows some new approaches found in the literature.

We have classified the approaches into four topics: ATPG techniques including timing issues, testing for crosstalk-induced delay, current testing applied to crosstalk and design for testability and on-line detection.

4.1 ATPG with timing implication considerations

One possibility of a new ATPG strategy is to consider analog and timing parameters, instead of only digital, as in the algorithm presented above [11]. From a lumped parameter capacitive model of crosstalk, spurious signals and induced delay are modeled with a piecewise linear model [12]. Using an 11-value algebra and adding signal arrival times, rise/fall times and input arrival skews, the ATPG tool explores all the possible spacings of pairs of input vectors using a modified backtrace algorithm. The objective is to generate a pair of vectors so the crosstalk effect is maximum. For example, in the case of crosstalk-induced added delay, given the timing of a clock-edge test generator, the test generator may generate tests that cause a victim line signal to slow-down and propagate the delayed signal in such a way as to violate the given timing requirement at a D input of a flip-flop.

Three objectives are considered in creating a very severe crosstalk effect: a weak driver on the victim line, a fast signal transition on the affecting line and a propagation path that maintains or amplifies the noise effect until it reaches an output. In order to evaluate this maximum noise created by the ATPG tool, a cost function is defined, which contains a digital part (controllability, observability measures) and an analog part (gate's capability to propagate noise). The backtrace procedure takes into account timing requirements, and from the information mentioned above including also path delay information, timing windows are obtained. These windows indicate time intervals in between which transitions must be made in order to create an important crosstalk effect. The ATPG tool can be applied to two different crosstalk effects, that is, crosstalk pulses and crosstalk induced delays. In both applications two different experiments are developed: firstly, if there is a fault, the tool can provide all the test vectors and select the best choice. Secondly, if there is a set of faults, it can provide a set of vectors for detecting these faults. Promising results are presented in this study [11].

4.2 Delay fault testing

Physical defects may not cause a logic error, but still cause problems by modifying the propagation delay of signals that cross the node where the defect is located. This behavior is modeled as a delay fault and several test techniques are devised to detect these faults. As crosstalk and SSN both produce a change in delay, the delay fault test techniques may, in principle, be applicable to noise testing and this is the reason they are briefly described here.

Similarly, as with the crosstalk test strategy described above, the delay fault test needs two vectors in order to generate a transition on which its propagation delay is somehow measured. The second input vector that generates the transition should preferably allow a single input to change, in order to avoid multiple transitions in the circuit that might mask the result. Transient measurements determine whether the observed transition at the output arrives inside or outside a given time window. In this scheme, the first vector defines the path sensitization where the fault is tested (first part of a controllability problem) and the second vector causes the transition and propagates it to one or more outputs.

Figure 6.7 shows an example of delay fault test application in a simple circuit. If a delay fault at path $A - B - C - D$ has to be detected, a first vector that sensitizes this path is (001001) (with inputs ordered from top to bottom). This vector enables the propagation of one transition from A to D by making all the gates of this path transparent. Next, a change in A at time t is forced, so the second vector is (001101) and the resulting transition time at output D is measured, which will occur at a certain time $t + T$. If the transition delay, T, is outside the margin defined by t_1 and t_2, then the circuit shows a delay fault.

In the case of crosstalk-induced delay, a similar technique can be applied by introducing slight changes. To start with, the first vector has to set the coupled nodes in the same way as for the general case, by sensitizing the path containing the victim line. The second vector is in the case of crosstalk delay

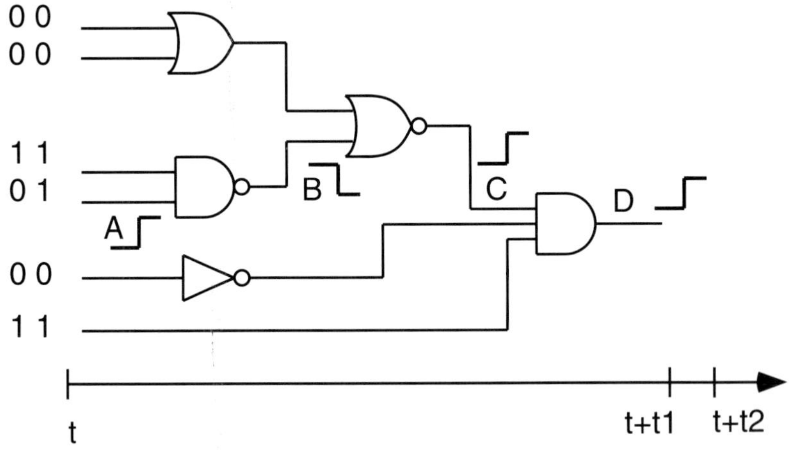

Figure 6.7. Circuit illustrating delay fault test.

test more complicated to find, because it has to force simultaneous transitions in the coupled nodes. The propagation delays from input to the coupled nodes will play an important role here, because if the transitions are not simultaneous the effect is very much reduced.

In [6] a test pattern generation technique is presented that includes the effect of SSN on the delay fault. The conclusions of the paper are that the inclusion of the SSN effects tends to significantly increase the circuit's delays.

4.3 I_{DDx} test approaches to crosstalk detection

The above-mentioned test techniques use the logic output of the circuit as the observable magnitude (and, more precisely, the timing of the logic signals). Another observable widely used for test purposes is the current consumption of the circuit. The static current consumption technique, or I_{DDQ}, is very efficient in detecting physical defects like bridges, opens, and GOS (gate oxide shorts in MOS transistors) [13], [14], [15], [16]. For defects that do not alter static current consumption, the dynamic current consumption can be used as the observable magnitude, as shown in recent works [17], [18], [19], [20], [21], [22], [23].

As shown in chapter 3, crosstalk-induced delay due to simultaneous transitions causes an increase in the energy consumed, which readily translates to an increase in current consumption. Spurious signals also cause an increase in current consumption because they represent extra voltage transitions in nodes that would otherwise be quiet. Therefore, a crosstalk problem implies a current consumption variation with respect to a circuit without crosstalk, as shown in figures 6.8 and 6.9. In the case of the circuit with a crosstalk problem, there

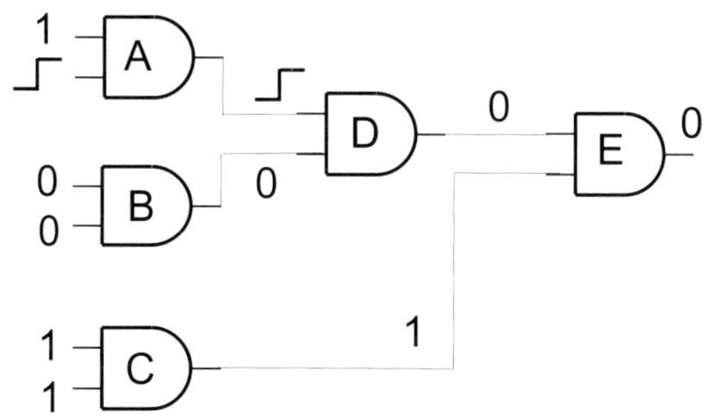

Figure 6.8. Circuit with no crosstalk problem.

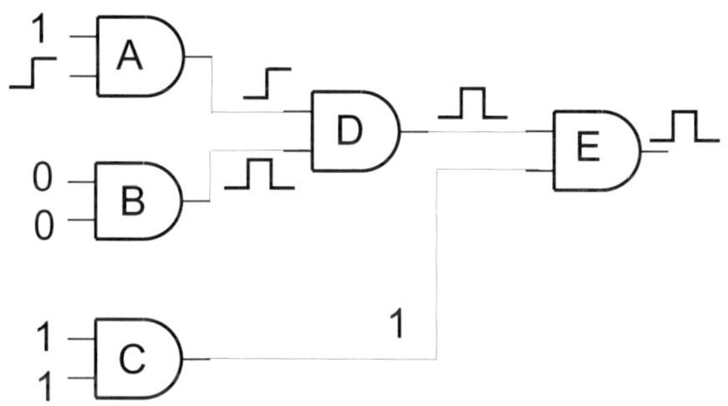

Figure 6.9. Circuit with crosstalk problem.

are more switching nodes, and its corresponding current waveform will be different.

A test strategy for crosstalk based on dynamic current consumption (I_{dd}) tries to cause the maximum crosstalk effect and the maximum propagation of the spurious signal through logic, so that the number of switching gates is at a maximum. In this strategy, whether or not the spurious signal arrives at some output is not so important, because the current is the magnitude observed.

In [24] the so called ECR (energy consumption ratio) [25] I_{dd} test technique is modified for detecting crosstalk. The resulting technique, called ECR-VDD, is applicable to both spurious signals and crosstalk induced delay. The ECR technique is based on the measure of the current consumption ratio of the circuit when applying two different input vectors. The ratio of average current consumption is defined as the ECR. The advantage of the technique is that, due to considering average currents, manufacturing process deviations to the test result (which is one of the main limitations of the I_{dd} in general) are minimized.

4.4 Design for testability and on line detection

Other aspects related to testing are design for testability and on-line crosstalk detection. Several works address these topics.

In [26] a self-testing methodology to enable on-chip, at-speed testing of crosstalk defects in system on-chip interconnects is described. The methodology is based on the Maximal Aggressor Fault Model [27] which enables testing of the interconnection with a linear number of test patterns. In this work, on-chip test generators and error detectors are designed in order to be embedded in the main cores. Test generators generate test vectors for crosstalk faults and error detectors analyze the transmission of the test sequences received from the interconnects, detecting any transmission errors. Test controllers manage the transactions and allows diagnosis capabilities. This technique has been applied to the buses of a DSP chip and, by inserting self-test structures, self-test capability has been obtained.

Finally in the area of self-testing and on-line testing, a recent work [28] shows a modification of a scan flip-flop that enables on-line detection of delay and crosstalk faults at the flip-flop's inputs. This type of circuit allows the circuit to have self- checking capabilities.

References

[1] A. Deutsch, H. Smith, G.A. Katopis, W.D. Becker, P.W. Coteus, C.W. Surovic, G.V. Kopcsay, B.J. Rubin, R.P. Dunne, T. Gallo, D.R. Knebel, B.L. Krauter, L.M. Terman, G.A. Sai-Halasz, and P.J. Reslte. "The importance of inductance and inductive coupling for on-chip wiring". In *Topical Meeting on Electrical performance of Electronic Packages*, 1997.

[2] T. Mido and K. Asada. "Crosstalk noise in high density and high speed interconnections due to inductive coupling". In *Asia-South Pacific Design Automation Conference*, 1997.

[3] Roth J.P., Bouricius W.G., and Schneider P.R. "Programmed algorithms to compute tests to detect and distinguish between failures in logic circuits". *IEEE Trans. on Electronic Computers*, EC-16(10), October 1967.

[4] Goel P. "An implicit enumeration algorithm to generate tests for combinational logic circuits". *IEEE Trans. on Computers*, C-30(3), March 1981.

[5] Fujiwara H. and Shimono T. "On the acceleration of test generation algorithms". *IEEE Trans. on Computers*, C-32(12), December 1983.

[6] Krstic A., Jiang Y. M., and Cheng K.T. "Pattern Generation for Delay Testing and Dynamic Timing Analysis Considering Power-Supply Noise Effects". *IEEE Transactions On Computer Aided Design of Integrated Circuits and Systems*, 16(3), March 2001.

[7] Moll F. and Rubio A. "Methodology of Detection of Spurious Signals in VLSI Circuits". In *Proc. of European Test Conference*, 1993.

[8] Moll F., Roca M., Marche D., and Rubio A. "Detectability of Spurious Signals with Limited Propagation in Combinational Circuits". In *IEEE 3rd Asian Test Symposium*, November 1994.

[9] Goldstein L. "Controlability/observability analysis of digital circuits". *IEEE Trans. on Circuits and Systems*, 1979.

[10] Brglez F. and Fujiwara H. "A neutral netlist of 10 combinational benchmark circuits and a target translator in Fortran". In *International Symposium on Circuits and Systems*, June 1985.

[11] Chen Wei-Yu, Gupta Sandeep K., and Breuer Melvin A. "Test Generation for Crosstalk-Induced Faults: Framework and Computational Results". *Journal of Electronic Testing: Theory and Applications*, 18(1), February 2002.

[12] Chen Wei-Yu, Gupta Sandeep K., and Breuer Melvin A. "Test Generation in VLSI Circuits for Crosstalk Noise". In *Proc. International Test Conference*, 1998.

[13] Rodriguez R., Pineda J., and Volf P. "Resistance characterization for weak open defects". *IEEE Design and Test of Computers*, September 2002.

[14] Roca M. and Rubio A. "Current Testability Analysis of Feedback Bridging Faults in CMOS circuits". *IEEE Transactions on Computer Aided Design of Integrated Circuits and Systems*, 14(10), October 1995.

[15] Suehle J. W. "Ultrathin gate oxide reliability: Physical models, statistics, and characterization". *IEEE Transactions On Electron Devices*, 49(6), June 2002.

[16] Segura J. and Rubio A. "A detailed analysis of CMOS SRAM's with gate oxide short defects". *IEEE Journal of Solid State Circuits*, 32(10), October 1997.

[17] Beasley J., Righter A.W., Apodaca C.J., Pour-Mozafari S., and Huggett D. "Idd pulse response testing applied to complex CMOS ICs". In *Proc. International Test Conference*, 1997.

[18] Makki R., Shyang-Tai S., and Nagle T. "Transient power supply current testing of digital CMOS circuits". In *Proc. International Test Conference*, 1995.

[19] Sachdev M., Janssen P., and Zieren V. "Defect detection with transient current testing and its potential for deep sub-micron CMOS ICs". In *Proc. International Test Conference*, 1998.

[20] Segura J., De Paul I., Roca M., Isern E., and Hawkins C. "Experimental analysis of transient current testing based on charge observation". *IEE Electronic Letters*, 35(6), March 1999.

[21] Picos R., Roca M., Isern E., and Garcia Moreno E. "Experimental results on BIC sensors for transient current testing". *Journal of Electronic Testing: Theory and Applications*, 16(3), 2000.

[22] Singh A., Plusquellic J., and Gattiker A. "Power supply transient signal analysis under real process and hardware models". In *IEEE VLSI Test Symposium*, 2002.

[23] Jiang W. and Vinnakota B. "Statistical threshold formulation for dynamic Idd test". *IEEE Transactions On Computer Aided Design of Integrated Circuits and Systems*, 21(6), June 2002.

[24] Sun X., Kim S., and Vinnakota B. "Crosstalk Fault Detection by Dynamic Idd". In *Proc. of the International Conference on Computer Aided Design ICCAD*, 2001.

[25] Jiang W. and Vinnakota B. "IC Test with the Energy Consumption Ratio". In *Proc. of the Design Automation Conference*, 1999.

[26] Bay X., Dey S., and Rajski J. "Self-Test Methodology for At-Speed Test of Crosstalk in Chip Interconnects". In *Proc. of the Design Automation Conference*, 2000.

[27] M. Cuviello, S. Dey, Xiaoliang Bai, and Yi Zhao. "Fault modeling and simulation for crosstalk in system-on-chip interconnects". In *International Conf. on Computer Aided Design (ICCAD)*, 1999.

[28] Metra C., Di Francescantoni S., Favalli M., and Ricco B. "Scan flip-flops with on-line testing ability with respec to input delay and crosstalk faults". *Microelectronics Journal*, 34, 2003.

Index

About the Authors

Dr. Francesc Moll Echeto obtained his PhD degree in 1995 by the Universitat Politecnica de Catalunya in the area of IC interconnect modeling and analysis. He is currently a professor in the Department of Electronic Engineering of the Universitat Politecnica de Catalunya. During 1998 he was in a research stay at the Dept. of Electrical Engineering of the State University of New York at Binghamton, using electromagnetic modeling tools for electronic package analysis. His research interests include VLSI physical design, noise analysis and noise avoidance techniques.

Dr. Miquel Roca Adrover obtained the PhD degree in 1994 by the Universitat de les Illes Balears. He was with the Department of Electronic Engineering of the Universitat Politecnica de Catalunya in Barcelona between 1991 and 1993. During 1996 he was in a research stay at the Departement de Genie Electrique et Informatique of the INSA in Toulouse (France). He is currently a professor at the Department of Physics (Electronic Technology Area) in the Universitat Illes Balears (Spain). His research interests include VLSI design and test, high speed circuits and crosstalk problems analysis in VLSI Circuits.